SERVICE CORPS

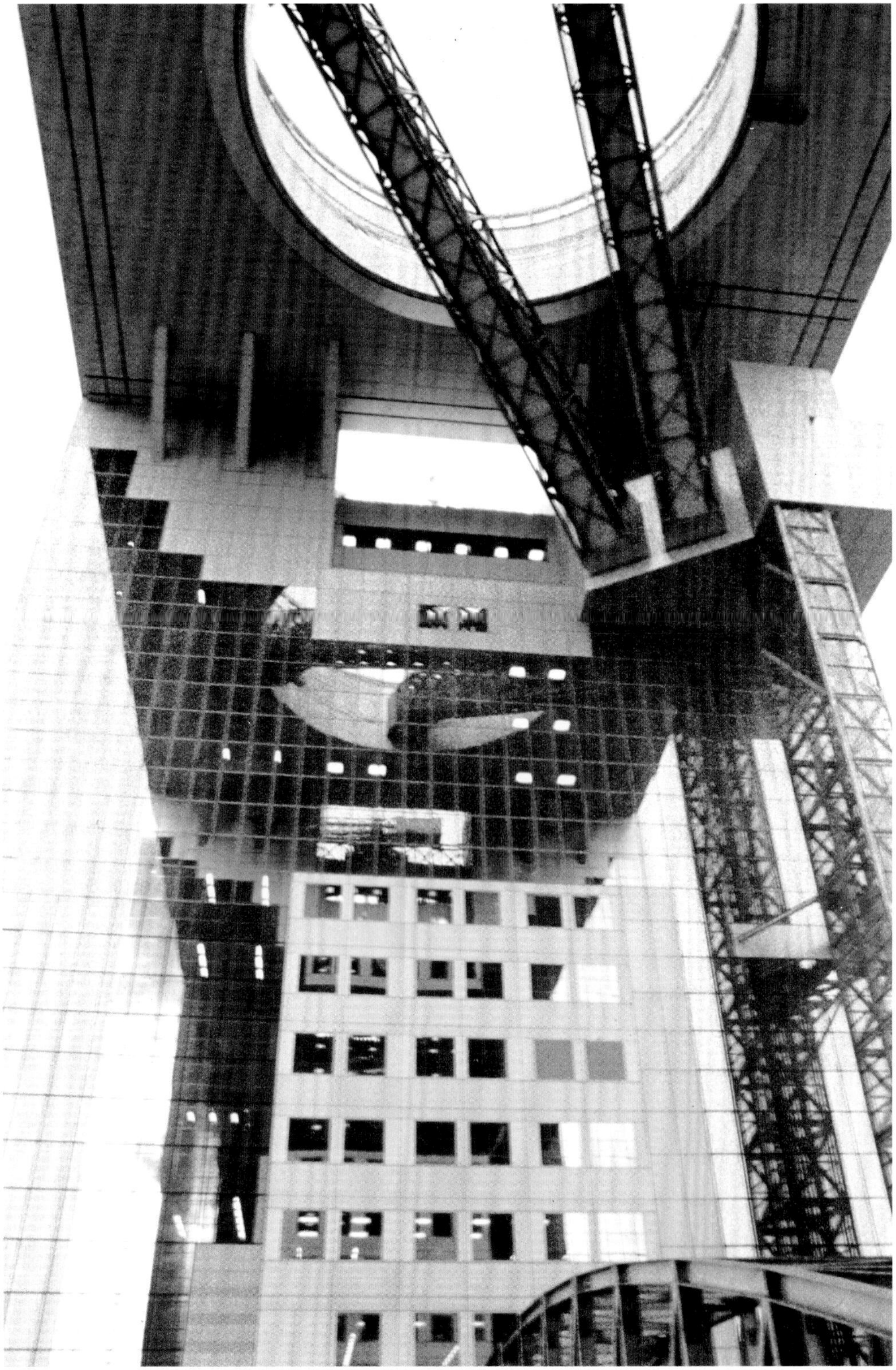

SERVICE CORES

KEN YEANG

DETAIL IN BUILDING

WILEY-ACADEMY

DETAIL IN BUILDING

Advisory Panel: Maritz Vandenberg, Christopher Dean, Christopher McCarthy, Maggie Toy, Michael Spens

Picture Credits
Every attempt has been made to locate sources and credit material but in the very few cases where this has not been possible our apologies are offered. All technical diagrams were drawn by the author; these are meant to convey principles and should not be read as precise working drawings. All other drawings and photographs are courtesy of the architects, with the exception of the following: p. 51, Nigel Young/Foster and Partners; pp.54 & 55 (top left), Kokyu Miwa; p.57, Kaneaki Monma; p.58, Mituo Matuoka; p.59, SS; pp.66 & 67 (top left), Georges Fessy; p.68, Michel Moch; pp.70 & 71 (top left), Tom Miller/Richard Davies; p.72, Nigel Young/Foster and Partners; p.73 (top left), Mishima/Foster and Partners; pp.74 & 75 (top left), K. L. Ng Photography; pp.76, 77 (top left) & 78 (top left), Tomio Ohashi; p.79, Kazuo Natori; p.80, Richard Einzig/Arcaid; p.82 (left), John Donat; p.82 (right), Peter Cook.

Cover: Foster and Partners, Commerzbank Headquarters, Frankfurt, interior view of atrium
Page 2: Hiroshi Hara, Umeda Sky Building, Osaka

First published in Great Britain in 2000 by
WILEY-ACADEMY

A division of
JOHN WILEY & SONS
Baffins Lane
Chichester
West Sussex PO19 1UD

ISBN: 0 471 97904 X

Copyright © 2000 John Wiley & Sons Ltd. *All rights reserved.*
No part of this publication may be reproduced, stored in a retrieval system, or transmitted, in any form or by any means, electronic, mechanical, photocopying, recording, scanning or otherwise, except under the terms of the Copyright, Designs and Patents Act 1988 or under the terms of a licence issued by the Copyright Licensing Agency, 90 Tottenham Court Road, London, UK, W1P 9HE, without the permission in writing of the publisher.

Other Wiley Editorial Offices
New York • Weinheim • Brisbane • Singapore • Toronto

Printed and bound in Italy

CONTENTS

Foreword	7
Introduction	**9**
Definitions	10
Design Approach	12
Function of the Service Core	13
Service Core Types and Placements	15
The Service Core and Building Economy	21
Structural and Construction Aspects	22
M&E Services	27
Opportunities for Optimisation	31
Net and Gross Area Definitions	32
Floor-Plate Configuration	34
Workspace Design	41
Elevator Shaft Configuration	42
Elevator Shafts Within the Service Core	45
Staircases, Exits and Life-Safety Considerations	47
Toilets in Service Cores	49
M&E Service Risers and On-Floor Plant Rooms	52
Conclusion	53
Case Studies	**54**
Nikken Sekkei, Shinjuku 'NS' Building, Tokyo	54
Nikken Sekkei, IBM Headquarters Building, Tokyo	56
Nikken Sekkei, Shinjuku Sumitomo Building, Tokyo	58
Nikken Sekkei, Mitsui Marine & Fire Insurance, Nagoya	59
SOM, Texas Commerce Tower, Dallas, Texas	60
SOM, Lever House, New York	62
Mitsubishi Estate, The Landmark Tower, Yokohama	63
Hiroshi Hara, Umeda Sky Building, Osaka	64
Jean Nouvel, Tour Sans Fins	66
Oscar Niemeyer, Congress Hall, Brasilia	68
Foster and Partners, Millennium Tower, London	70
Foster and Partners, Commerzbank HQ, Frankfurt	72
Foster and Partners, Century Tower, Tokyo	73
Ken Yeang, Menara TA1, Kuala Lumpur	74
Kisho Kurokawa, Melbourne Central, Melbourne	76
Kisho Kurokawa, Nakagin Capsule Tower, Tokyo	78
Nihon Sekkei, Jing Guang Centre, Beijing	79
J Stirling & J Gowan, Leicester University Engineering Building, Leicester	80
Ben van Berkel, Scuba Tower, London	81
Richard Rogers, Lloyd's Building, London	82
Kohn Pedersen Fox, 333 Wacker Drive, Chicago, Illinois	84
Cesar Pelli, Petronas Towers, Kuala Lumpur	86
Checklist of M&E Services (for integration with service cores)	88

Cesar Pelli, Petronas Towers, Kuala Lumpur

FOREWORD

As the *Detail in Building* series has developed through the years it has become clear that each subject featured requires slightly different treatment. Whilst the common intention of exploring and exposing a particular building detail remains the same the nature of the presentation is altered to suit the subject area.

From its origins in the nineteenth-century American high-rise office block, the skyscraper has become the dominant building type of the twentieth century. Developers have been quick to seize upon its potential to maximise land use, and rapid advances in technology and construction techniques have led to a proliferation of ever taller buildings. In this publication Ken Yeang focusses on one of the most essential elements in this ever elevating building type, the service core, and demonstrates how refining its design can allow considerably more freedom and finesse in the overall building itself.

Based on the research, design and development of the author's own architectural practice in Kuala Lumpur, Malaysia, the text is illustrated not only by an extensive and wide range of tall buildings world wide from past to present, but also with examples of his own work. He demonstrates that new techniques are necessary to take this building type to its inevitable next stage; a more eco-friendly building is required and as Yeang shows is now a reality.

Professor Ivor Richards explains: 'As the old technology reaches its limits, the age of the super-skyscraper dawns, driven by emergent new forms and new technologies – the ultimate building becomes a city in itself.'

Maggie Toy

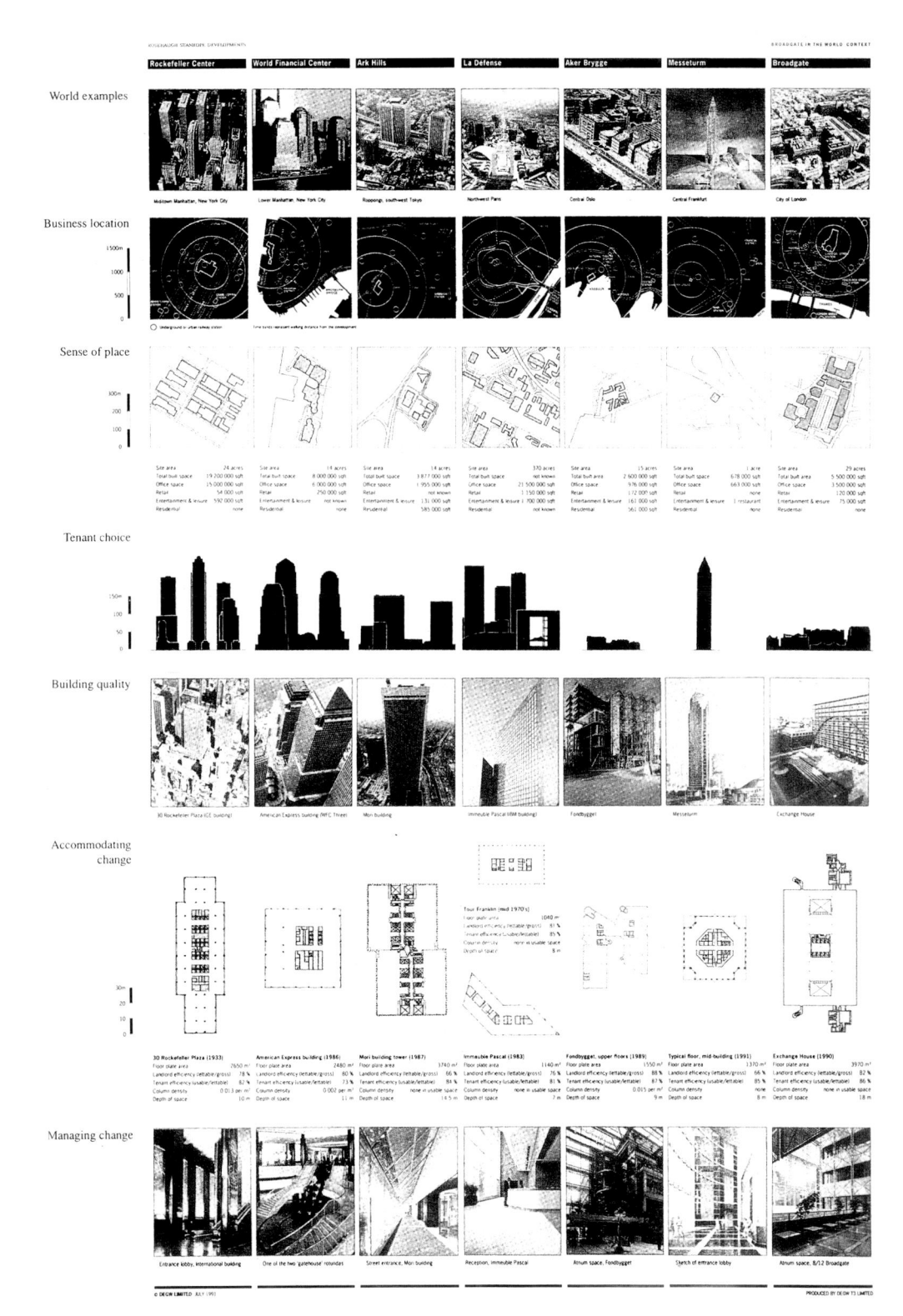

Fig. 1

INTRODUCTION

Service cores are an increasingly important aspect of building design, architecturally, structurally and from the building scientist's viewpoint. This importance is attributive to many factors, not least of which is the increased level of mechanical and electrical (M&E) engineering systems within modern buildings. The result, though, has been a general increase in size of the service core. The knock-on effect has been to alter the building's overall space efficiency or, to the real estate agent, the building's net-to-gross area ratio. Some designers even regard the core as the most important part of the building.

The service core in tall buildings and skyscrapers may become more crucial as designers push the envelope of construction into the sky (Fig. 1). In these building-types the structural engineer, along with the vertical transport engineer, will often have the most significant role in the design since the core will be multifunctional in its use. In highly serviced buildings such as those for the pharmaceutical industry or hospitals, the M&E engineering systems will dictate the size and location of the core. In office and low-energy buildings, the architectural concept tends to be the driving force in the core strategy.

Users of all the various types of building regard the service cores as spaces away from the working environment and meeting places (Fig. 2), whereas the letting agent, and often the owner, may see them as necessary but redundant spaces.

When making initial design decisions, the placement of service cores within the floor plate can significantly affect the M&E system's distribution routes, as well as the vertical circulation system and the efficiency of use of the building (Fig. 3).

The following has been written as a guide for the designer to follow. It is not prescriptive, but should be treated as an aide-mémoire. Each and every building is unique in its own way and, for no lesser a reason, the design of the cores should be so too.

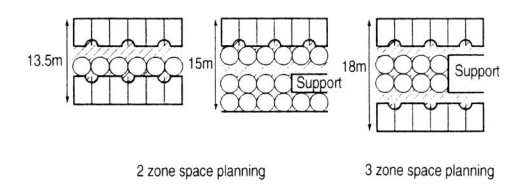

Fig. 2

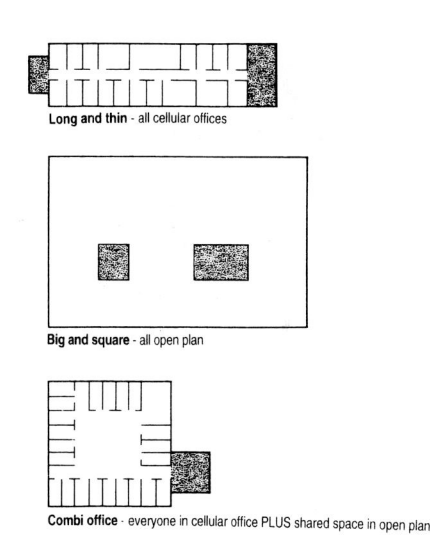

Fig. 3   Three Types of Buildings: long and thin, big and square and a Combi office

1. DEFINITIONS

Simply stated, a service core is defined as those parts of a building that consist of the elevators, the elevator shafts, the elevator lobby, staircases, toilets, M&E service, riser ducts and, in some cases, the M&E plant rooms. Its structure can also contribute to the structural stability of the building.

Service cores can typically contain the following elements:

- elevator shafts (inclusive of the elevator cars and equipment inside them)
- elevator lobby (into which the elevator shafts open)
- staircases (usually consisting of a main staircase and an escape staircase)
- fire-protected lobbies (where these are required, depending on the configuration and level of fire protection, building type and size)
- toilets (which usually consist of male and female toilets, disabled persons toilets, and executive toilets, where provided)
- ancillary rooms such as pantry, space for cleaning materials, where these are provided
- mechanical vertical services riser-ducts, e.g. for electrical power and lighting distribution, water distribution (including both riser and dropper pipes), sewerage pipes, rainwater downpipes, system-medium piping, hot water piping, fire-fighting pipes and equipment, exhaust ducts, etc
- structural bracing and stability, achieved in the elevator shaft and staircase design (if applicable),
- mechanical vertical fire-protection risers for sprinklers, hose reels, wet and dry risers
- electrical vertical service riser for power
- electrical vertical service risers for telecommunications and data systems
- M&E services plant rooms (where required for air handling units, telecommunications-distribution equipment, etc
- Walls (to the service core) which can contribute to the structural stiffness of the building

One approach to designing a highly serviced building whose function is dependent on the service core is to analyse by design the implications of different positions for the service core within the building's floor plate. The position of the service core in relation to the usable areas in the floor plate essentially determines the vertical circulation system of the building and how most of the building's M&E services are distributed. An example of the contents of a service core, in this case a split service core, is shown in Fig. 4.

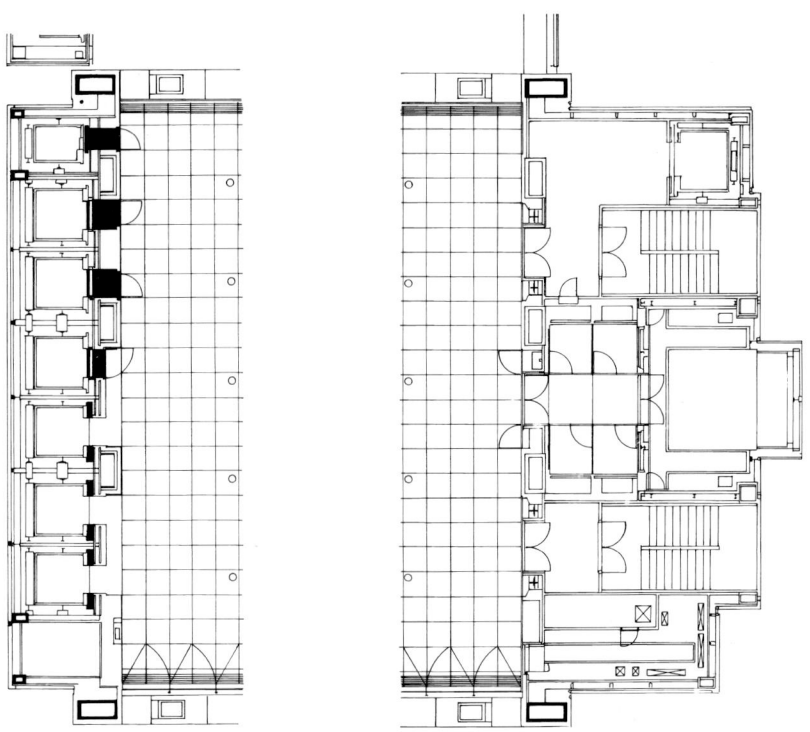

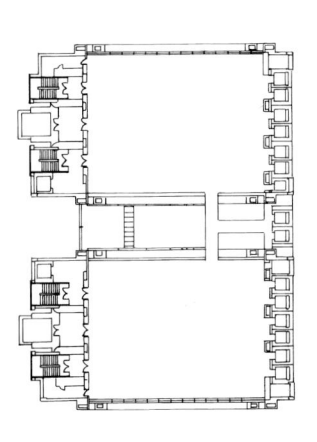

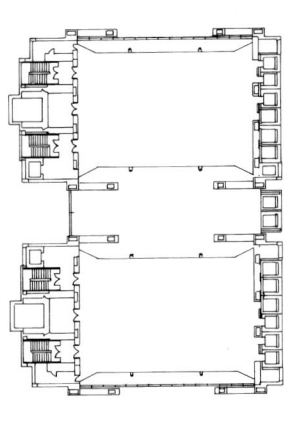

Fig. 4: Plan and core detail of end-core layout

2. DESIGN APPROACH

At the concept stage the design team should consider the implications and ramifications of all the sensible core placement options available. The major aspects which require addressing are the architectural design intent in the brief, functionality of the spaces, fire escape regulations, overall structural stability, M&E servicing, building topology and cost.

Smart core design will possibly play a key role in the office buildings of the 21st century. With the establishment of the electronic office environment and cordless technology, the office of the future may be a combination of a university – a place in which to learn and to share knowledge – and a hotel, used temporarily and with various shared facilities. The rapid development of cities like Frankfurt, London, New York and Tokyo has been a reason for the revival of high-rise office buildings which are beginning to proliferate on their skylines. Perhaps the office of the future will be arrangements of non-assigned workstations or 'hot' desks with more space given to communal facilities to allow the exchange of information both verbally and electronically.

3. FUNCTION OF THE SERVICE CORE

The size and location of the service core is predominantly governed by considerations that include the fundamental requirements of meeting fire-egress regulations, achieving basic efficiency in human movement and creating an efficient internal layout to maximise returns, and satisfy the requirements of vertical transport and the numerous vertical service shafts (Figs. 5A, 5B).

In tall buildings in particular, the service core can provide the principal structural element for both the gravity load-resisting system and lateral load-resisting system, with the latter becoming increasingly important as the height of the building increases. It provides the stiffness to restrict deflections and accelerations to acceptable levels at the top of the building.

Fire resistance of the core elements can be determined from the appropriate building regulations. It is noted that all penetrations for services through the walls of the service core need to be designed to maintain fire integrity for the prescribed period of time.

The core configuration is normally finalised at an early stage of the design development because of its impact on the functional layout of the building. Traditionally, the configuration is greatly influenced by the architect. The design optimisation process is subsequently carried out within the allocated zones during the preliminary design phase by the design team's experts in the individual disciplines.

The cost of a core for a typical high-rise office building is estimated to be around 38 per cent of the total structural cost or at 4 per cent to 5 per cent of the total development cost. Clearly, if the optimisation of the service core by the structural engineer is limited only to structural optimisation, the potential savings in terms of the overall cost is relatively small. In contrast, an optimisation process on the building's structure which impacts on the costs of other systems and which takes into account the speed of construction may result in more significant financial benefits.

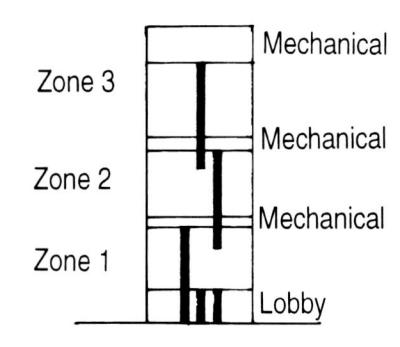

Fig. 5A: Diagram of elevators (three zone system in which users have to change floors after each zone)

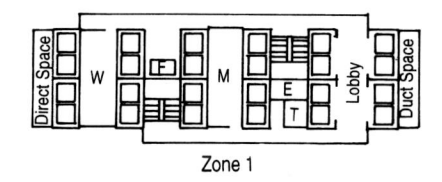

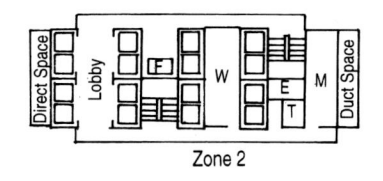

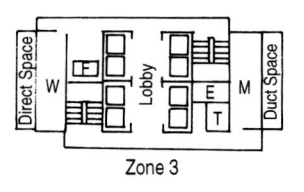

Fig. 5B

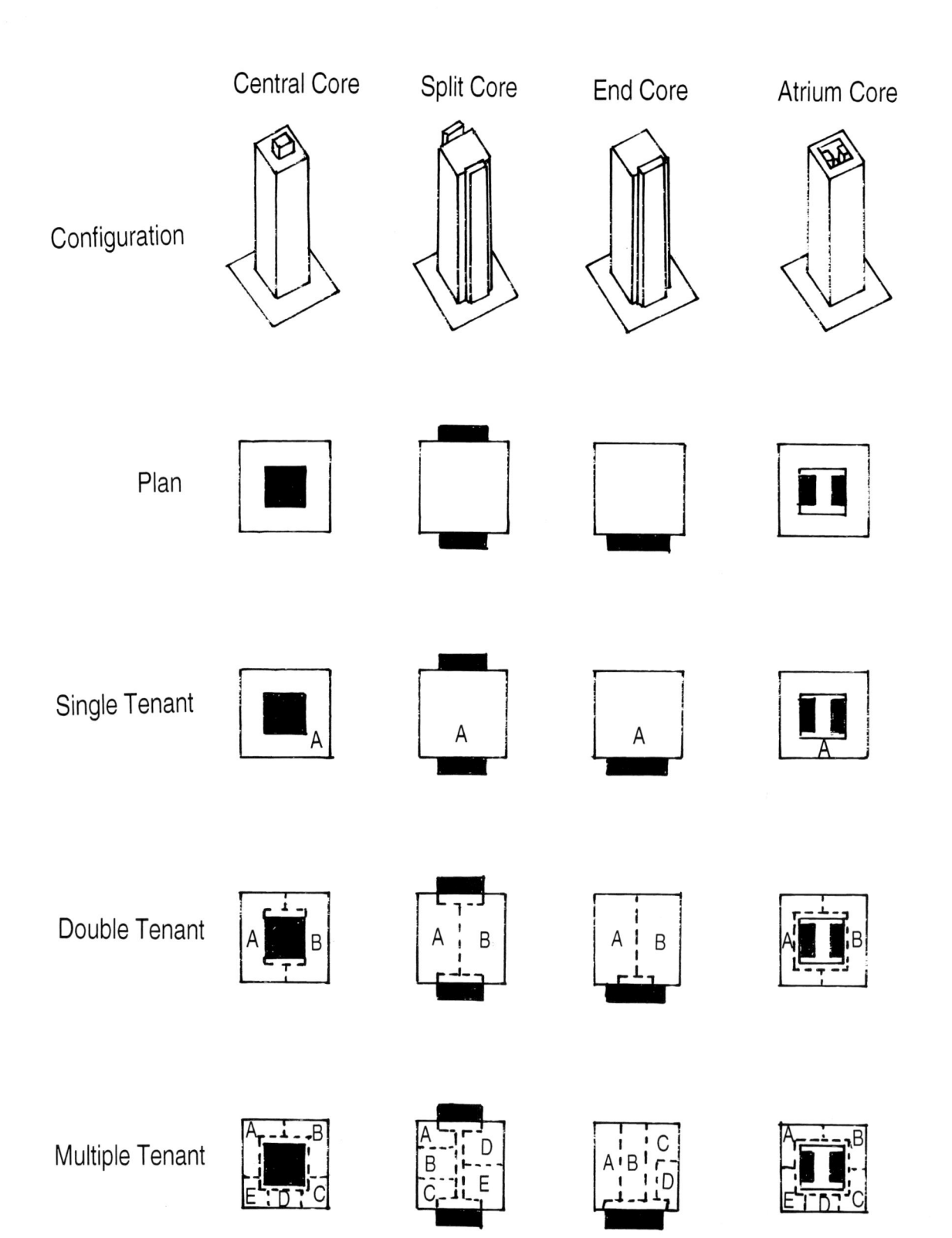

Fig. 6A

4. SERVICE CORE TYPES AND PLACEMENTS

How do we go about deciding where to place the service core when shaping and organising the plan of an office building or a highly serviced building?

The placement of the service core stems from four generic types which are used to design floor plates that meet the spatial requirements of the brief. They are known as:
- the central core
- the split core
- the end core
- the atrium core

Fig. 6A shows the generic arrangements and Fig. 6B shows the hybrid possibilities.

The approach will depend on many factors – hence the need to involve the whole design team. Selecting the correct criteria for the core design will help find the solution that will best meet the building's objectives. For instance, if unencumbered clear internal space is one of the main aims of the design, a single-ended core may be the best solution although the permissible distance from the furthest corner of the space to the fire escape is a limitation. A building with a lower energy brief may require a split-core configuration, which provides the best passive low-energy performance in high-rise buildings. Here the atrium should have a minimum width of 12 metres.

The placement as well as the internal arrangement of the core – the relationship of the core elements – depends very much on the type of building, the people who use the spaces and on legislation and building codes. The flexibility required by a multitenant user is quite different to what is needed in an owner-occupied building (see the lower diagrams in Figs. 6A and 6B). Certain elements will not change much. Elevators, staircases, toilets and service ducts are required in both. The flexibility in a speculative type of building is in the M&E engineering systems. The ability to carry out bespoke fit-outs from a 'shell and core' construction provides the level of interchangeability within a set of guidelines. Generally stated, glass-to-glass depths of up to 13.5 metres for floor plates with slab-to-slab heights of 3.6 metres to 4 metres, or a glass-to-core depth of 6 metres to 12 metres with a slab height of 3.8 metres to 4.5 metres, are likely to provide the widest range of servicing and space-planning options (Fig. 7).

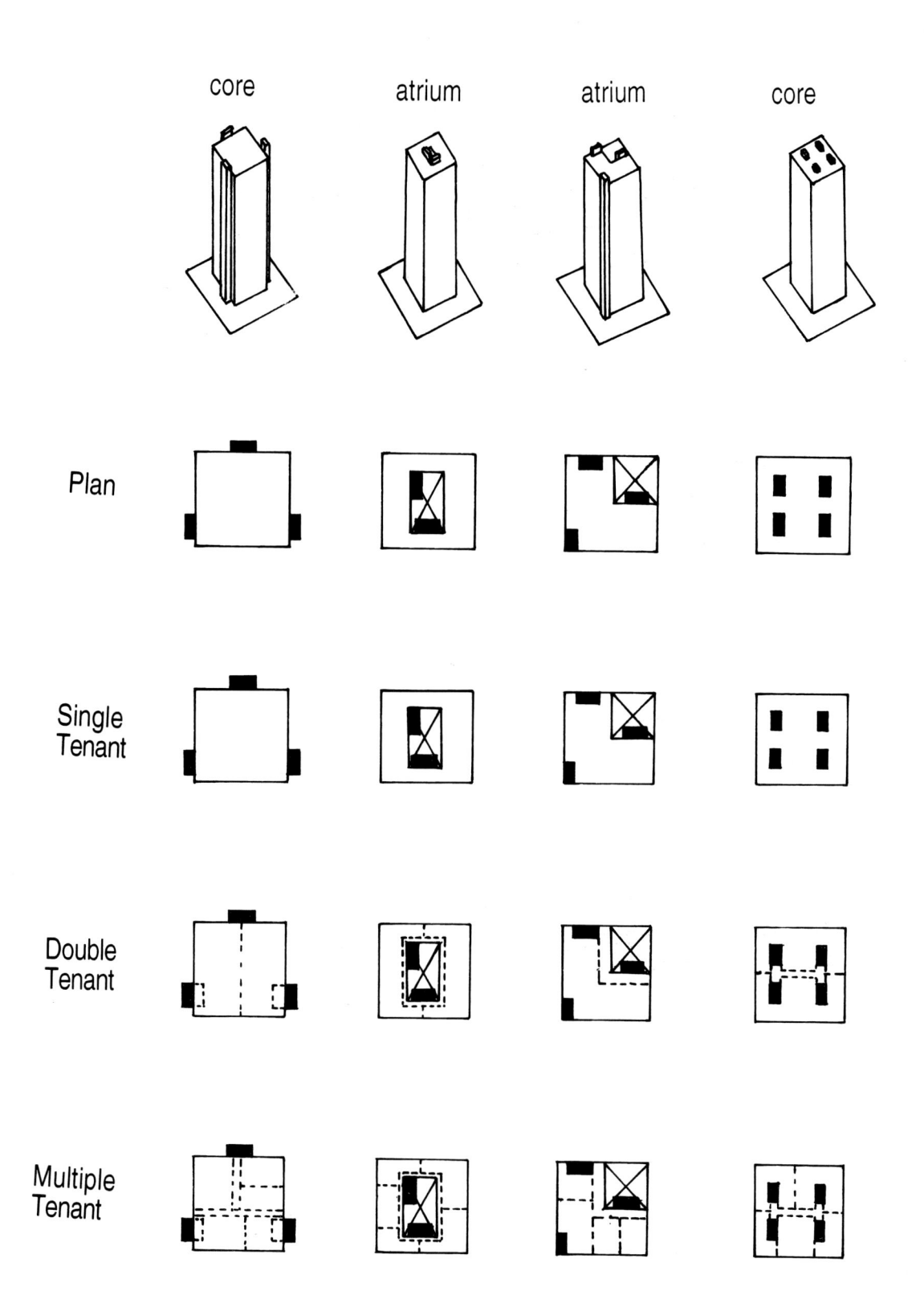

	core	atrium	atrium	core
Plan				
Single Tenant				
Double Tenant				
Multiple Tenant				

Fig. 6B

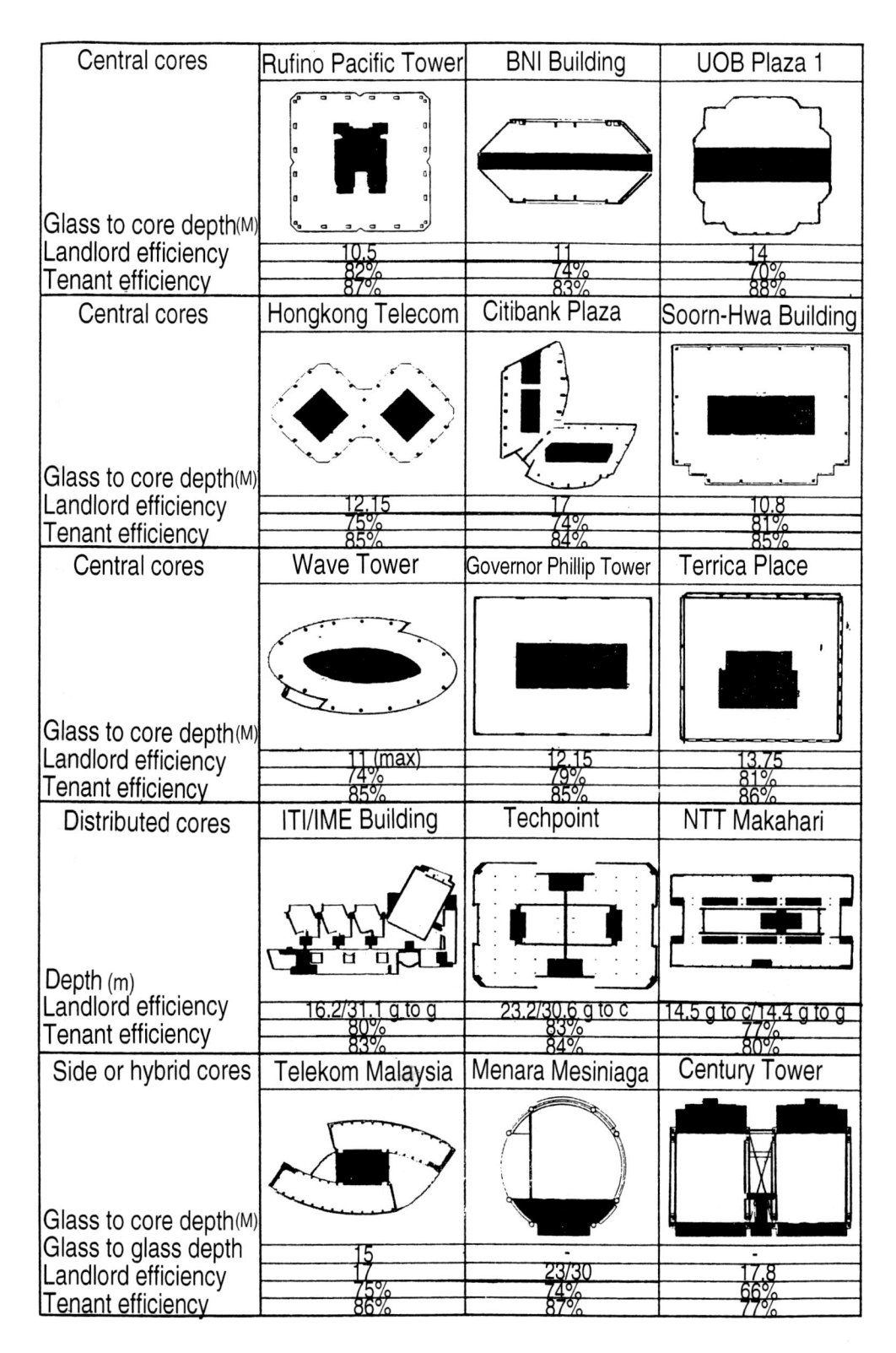

Central cores	Rufino Pacific Tower	BNI Building	UOB Plaza 1
Glass to core depth (M)	10.5	11	14
Landlord efficiency	82%	74%	70%
Tenant efficiency	87%	83%	88%

Central cores	Hongkong Telecom	Citibank Plaza	Soorn-Hwa Building
Glass to core depth (M)	12.15	17	10.8
Landlord efficiency	75%	74%	81%
Tenant efficiency	85%	84%	85%

Central cores	Wave Tower	Governor Phillip Tower	Terrica Place
Glass to core depth (M)	11 (max)	12.15	13.75
Landlord efficiency	74%	79%	81%
Tenant efficiency	85%	85%	86%

Distributed cores	ITI/IME Building	Techpoint	NTT Makahari
Depth (m)	16.2/31.1 g to g	23.2/30.6 g to c	14.5 g to c/14.4 g to g
Landlord efficiency	80%	83%	77%
Tenant efficiency	83%	84%	80%

Side or hybrid cores	Telekom Malaysia	Menara Mesiniaga	Century Tower
Glass to core depth (M)	-	-	-
Glass to glass depth	15	23/30	17.8
Landlord efficiency	75%	74%	66%
Tenant efficiency	86%	87%	77%

Fig. 7

For instance, if the building is entirely owner occupied, and if the occupier is not likely to make major internal changes in occupancy in the future, the building can be tailored entirely to meet the owner's unique requirements. An owner-occupied building can be more interesting to the design team since the guidelines are written by the owner and designers. Understanding the client's business will allow tailored and unique solutions to be developed. Floor plates can incorporate features such as irregular shapes, longer depths from the unlit end to the naturally sunlit external wall, the choice (or nonchoice) of ceiling grids, etc.

If you take a speculative office building as an example, the designer will need to establish the service core of the building around a variety of potential tenants. This flexibility will allow the eventual occupiers a full range of options when they subdivide the lettable space. Future refurbishments will be also easy when extending the useful life of the building. Generic speculative building-types have to cater for single or double tenants and, at worst, multitenants (Figs. 6A and 6B). Each requires different net-to-gross floor area efficiencies. The most efficient should be the single-tenancy occupancy. Emergency escape requirements, and hence the fire protection requirements, will be different. This will affect the design of the elevator lobby, which itself imposes on the elevator strategy. This may result in a structural solution which requires more lateral stability than can be achieved in the core.

However, as is often the case, the building (say, in this instance, that it is an office building) may be built by an investor who may not have any immediate occupant, or may have an occupant for only a portion of the building and not its entirety. In such instances, when the designer is planning the floor plate and placement of the service core, he or she should facilitate a variety of alternative options to allow for the full range of likely subdivisions of the internal spaces in the future.

The designer should not be blinkered into accepting the norms and traditions of core and floor-plate design. Some designers find planning the floor plate too constraining. His or her flair will be shown in their being able to deliver a nonrectilinear space, while meeting the strict code requirements for the safety of the occupants and satisfying of the accountants.

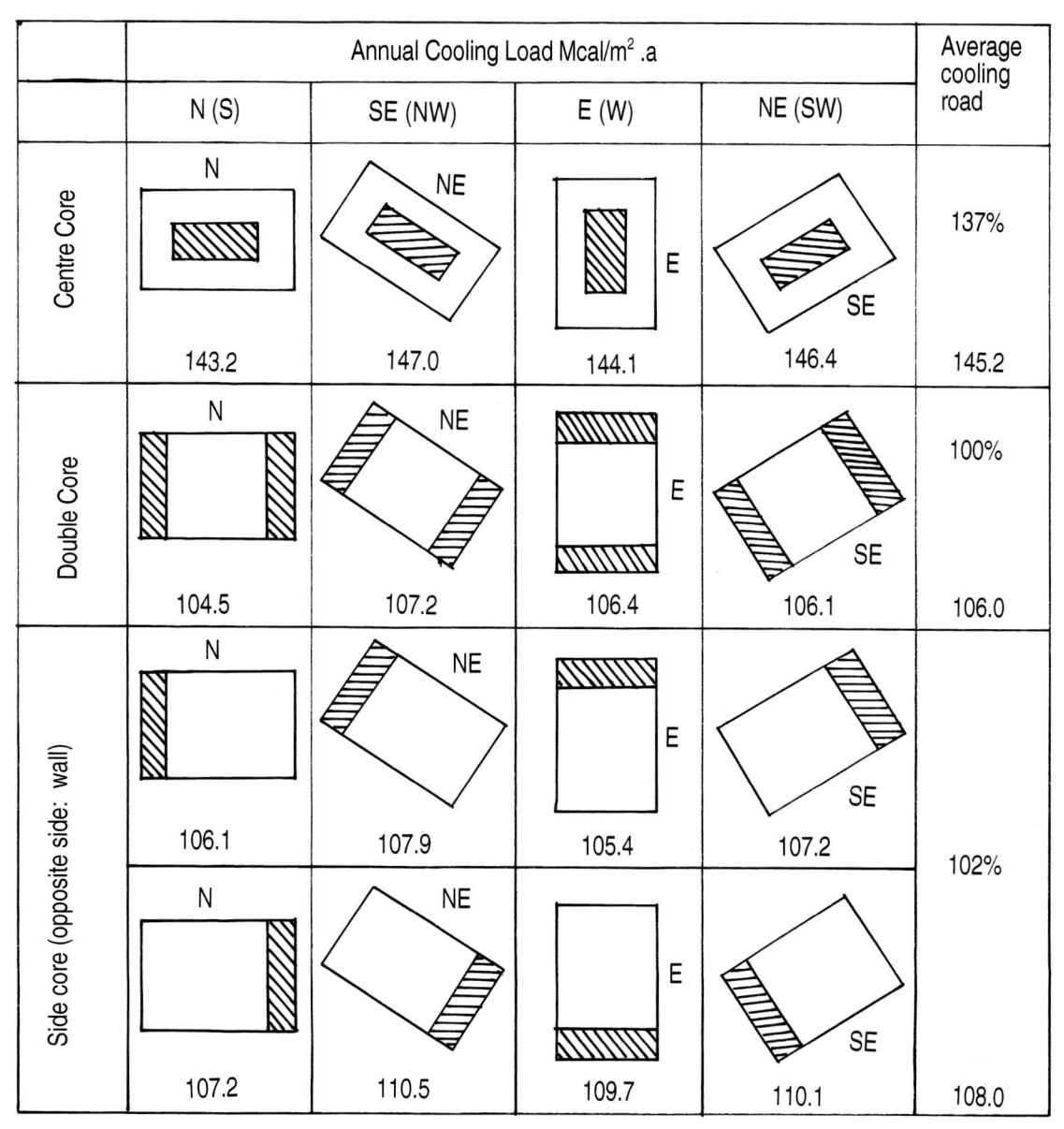

	Annual Cooling Load Mcal/m² .a				Average cooling road
	N (S)	SE (NW)	E (W)	NE (SW)	
Centre Core	143.2	147.0	144.1	146.4	137% 145.2
Double Core	104.5	107.2	106.4	106.1	100% 106.0
Side core (opposite side: wall)	106.1	107.9	105.4	107.2	102% 108.0
Side core (opposite side: wall)	107.2	110.5	109.7	110.1	

Notes:

Location	: Tokyo (Latitude 36ºN)	Temp./humid	
Typical floor area	: 2,400 m²	Cooling	: 26ºC 50
Floor height	: 3.7m	Heating	: 22ºC 50
Window-wall ratio	: 60	Air cond. floor area ratio	: 65
Lighting	: 30W/m²	Outdoor air intake	: 4.5m³/m²h
Infiltration	: 1 change/h	Length-breadth	: 1 : 1.5
People	: 7m² / p	Ratio insulation	: Foam polysthyrene 25mm

Fig. 8

The energy consumption of a building is greatly affected by the placement of its service cores. It will, of course, also depend on many other factors including geographical location and local topography as well as the type of building. There is a correlation between the location of the service core and the heating and cooling loads of the building. The cooling load is most influenced by the service core position. It is not surprising that a split-core design with the cores orientated east – west, with glazing to the north and south, has a lower cooling load than a central-core design. The cores on the east and west elevations reduce the high solar gain – usually the main constituent of the cooling load – to the building. Conversely, the service core placement that is characterised by the maximum air-conditioning load is the centre-core configuration in which the main daylight openings lie to the southeast and northwest. Fig. 8 shows the comparison of cooling load to core type. A building with a large lobby can suffer from chronic infiltration of ambient air throughout its interior, distributed via the elevator shafts and staircase. The results of neglecting the correct infiltration can be HVAC systems, which cannot maintain internal design conditions.

In the case of a skyscraper with a split-core design, a building orientated from north to south instead of east to west will have an air-conditioning load nearly one and a half times greater than a building arranged longitudinally from east to west. The service cores can be placed to serve as solar buffers, thereby enabling a passive low-energy bioclimatically responsive configuration. The need to cut down on electrical consumption and reduce carbon dioxide emissions makes good core design fundamental, not only environmentally but also financially.

There is also a trend today to have a greater proportion of the core area allocated to communications risers and IT facilities. In the past, designers placed these in cabinets along the walls of the service core. However, many are now providing these within the core area and enabling free wall-space around the core. There are, of course, spatial and efficiency penalties for doing so.

Concentrating all the M&E systems within the same zone achieves a better lettable area for the occupier and simplifies maintenance. Attention should be given to detailing the doors where risers open in protected and common access spaces. Coordination at the design stage will help provide a more user-friendly building. The designer should ensure that no common M&E service cabinets, plant rooms or ducts (usually within the service core area) have doors that open on to a rentable, usable or tenant's area. This would prevent occupants placing fittings and fixtures against that part of the service core's wall. Furthermore, locating the doors within a tenant's area would inhibit inspection by service personnel.

5. THE SERVICE CORE AND BUILDING ECONOMY

Previous studies have shown that the principal factors that contribute to overall building economy are:

Minimisation of material costs

The designer's objective is to optimise the concrete strength, wall thickness and reinforcement for a given core geometry (traditional structural optimisation).

Optimisation of core geometry

The service core layout should be geometrically efficient if it is to resist the structural loads, whilst fulfilling its architectural and services functions.

Minimisation of core area

The structural designer must appreciate the capitalised value of decreasing the service core area. To illustrate the significance of the potential benefits, a modest reduction of, for example, 50 millimetres in the thickness of all walls of a concrete-walled core represents significant savings in costs: 30 per cent to 40 per cent of the total structural costs of the core (Fig. 9).

Minimisation of construction time

The service core construction is typically a critical activity for any tall building development. Delays will affect the completion of the project and, ultimately, the income from potential tenants or occupants and will incur interest costs on the total project.

The core geometry and design must suit the builder's preferred method of construction, which depends on the equipment and expertise that is available at the locality of a project. Although a builder is rarely chosen at an early stage of design development, the designer must be aware of the above factors.

The two common methods of constructing concrete cores are slip form and jump form (different methods of formwork). In both, initial establishment costs are high relative to the construction costs.

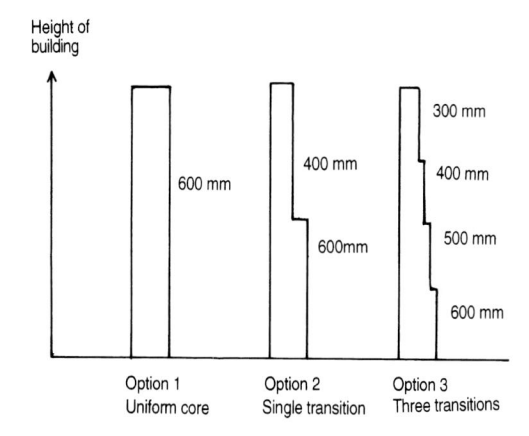

Fig. 9

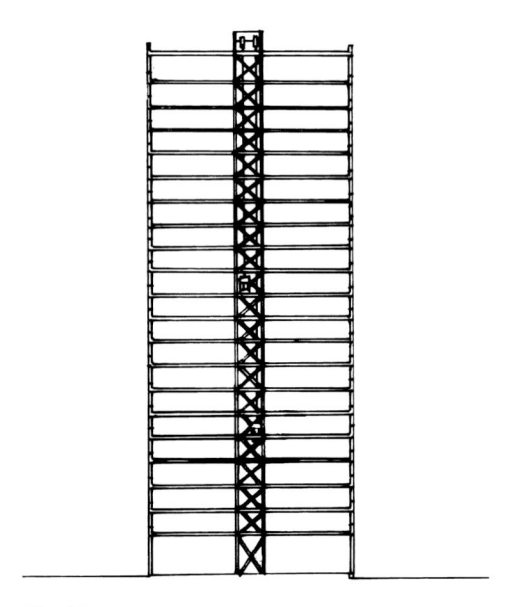

Fig. 10

6. STRUCTURAL AND CONSTRUCTION ASPECTS

For skyscrapers, the weight of the structural materials for the floor is generally the same whether the floor slab and its supporting beams are on the 10th or the 100th storey, since each floor carries more or less the same load. This is not the case with the core's columns, which carry the weights of all the floors above. The lower-floor columns carry much larger loads than those at the top. The topmost columns carry only the load of the roof and their own weight.

The load on the columns increases with the number of floors in the building, and their weight and load-bearing capacity must vary accordingly. However, both the wind forces and their lever arms increase with the height of the building. Their product, which measures the tendency to turn the building over increases with the square of the height. With the high strength achieved today in both steel and concrete structural construction, the swaying caused by the elasticity of these materials must be limited to ensure the comfort of the building's occupants. In earlier skyscrapers, horizontal rigidity was obtained by using closely spaced columns and deep beams, and by filling the voids between them with heavy masonry pavels such as brick walls.

The key to lighter frames lies in channelling the gravity loads to earth and the resistance to wind forces to two separate structural systems: relatively flexible exterior frames and a stiff, wind-bracing inner service core, inside which the elevators and vertical M&E pipes and ducts run. In addition to beams and columns, the service core can have a frame with diagonal bars that X the openings and give the core greatly increased stiffness by working in tension and compression rather than in bending (Fig. 10). The longitudinal deformations due to pulling and pushing are small compared to the lateral deformations due to bending. It is not difficult to see how the triangulated frames of the core can be light, but stiff enough to resist almost all the wind forces.

Diagonal bracing cannot be systematically used in exterior frames without cutting across windows, or in interior frames without destroying door space. Three of the four frames surrounding the core can be X-ed, and the fourth can be stiffened around the relatively small elevator door openings.

Structurally, the combination of two different materials could achieve the same results with increased economy. Rather than stiffening the core frames by means of diagonals, thin concrete walls can be used to give great stiffness in the horizontal direction. It then becomes logical to use reinforced concrete walls for all sides of the core, thus obtaining an interior narrow, self-supporting, stiff tower to which the outer frame, which may be of steel or concrete, is attached. In steel construction, 'rigid' or 'moment' (bolted or welded) connections are costly and require specialised manpower and dangerous work at great heights. The cost may represent 10 per cent of the entire cost of the structure. If the inner service core is stiff enough the need for the rigid connections between the beams and columns of the exterior frames can be reduced, and much cheaper connections which allow beams and columns to rotate one with respect to the other, as if they were hinged, can be used. Such hinged or 'shear' connections cannot be used without a stiff service core because the frames would collapse like a house of cards, but are economical and practical if the service core stands up rigidly and the outer hinged frame leans on it. The separation of the two structural functions is now complete: the hinged frame carries the vertical loads to the ground and the core resists the wind forces.

The principle of using the shear walls or X-ed inner-core steel frames of a concrete service core together with a hinged outer frame has reduced the weight of skyscrapers very considerably. We now frequently find that the walls of an elevator shaft are built from in situ concrete (for example, by means of slip forming) and are used structurally as shear walls to laterally stiffen the building's structure or enable the elevator shafts to be used as part of the system of vertical structural support columns.

However, for supertall skyscrapers (50 storeys and above), if the service-core and elevator-shaft walls are built in concrete their structural weight can become excessively heavy, thereby affecting the cost of the foundations. Some designers may choose to reduce the building's overall weight by eliminating concrete shear walls and using properly sealed fire-rated gypsum boards for service core walls, with the structural function transferred to the other columns and the lateral structural stiffening achieved by other means such as enlarged columns or horizontal cross-bracing.

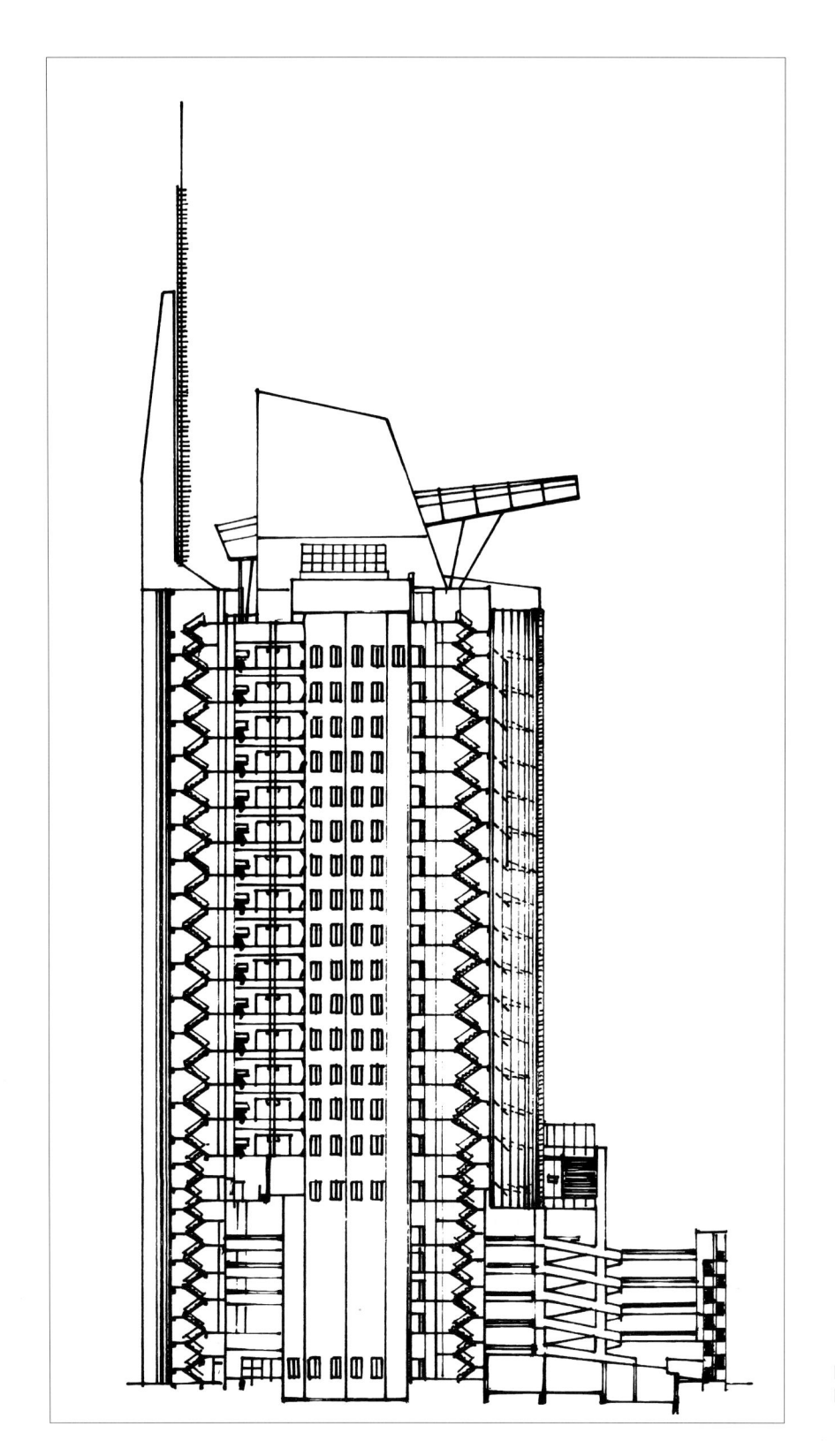

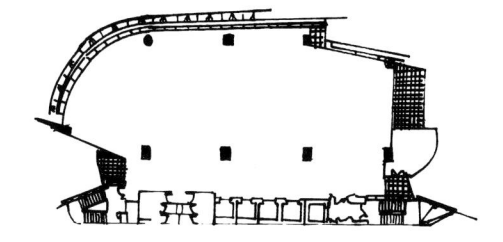

Fig. 11A-B: Menara UMNO, Yeang
Penang, Malaysia

24

The impact of service core construction on direct construction costs and indirect 'soft' costs associated with interest charges and returns offered on the lease of floor areas relate to the following factors:
- material costs,
- construction cost associated with core-wall thickness transitions,
- construction time associated with core-wall thickness transitions,
- capitalised value associated with the increase in net rentable area.

Issues affecting construction time and the financial benefits of increasing lettable space are the major contributing factors in determining the real cost of a core. The cost benefits of having transitions in the thickness of the core walls in order to maximise rentable space need to be assessed with extreme caution. The transitions cause delays that can dramatically offset the benefits – and, with inadequate knowledge of construction issues, can lead to an overall loss.

Clearly, decisions need to be made well before a builder is involved to any significant extent. This emphasises the importance of seeking good construction advice on core technology at this very early stage in a design if substantial savings are to be made.

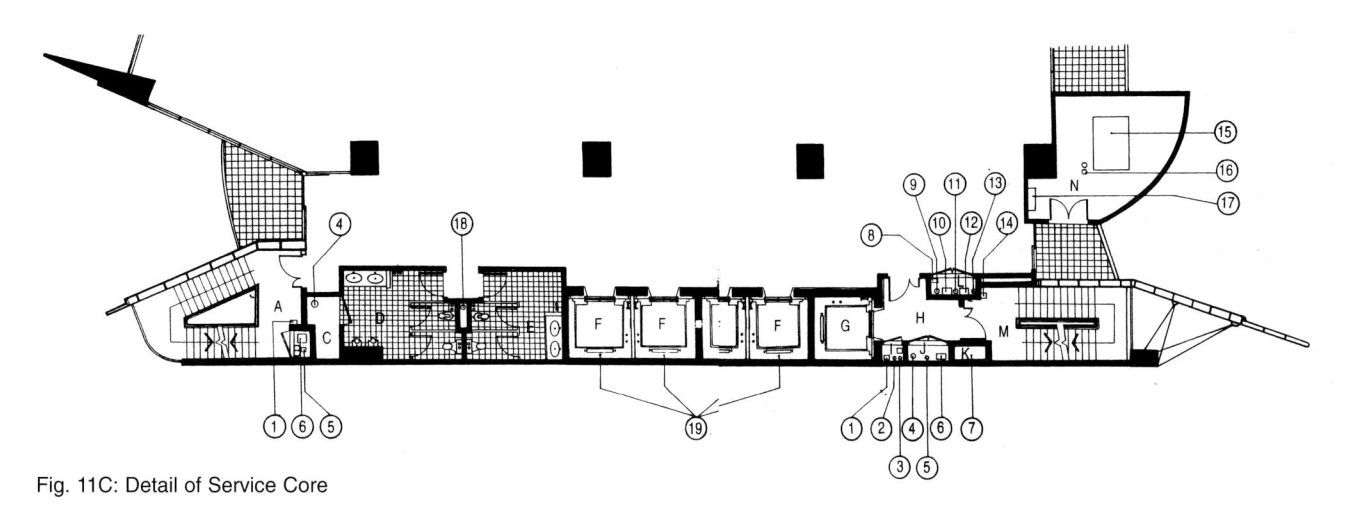

Fig. 11C: Detail of Service Core

Ken Yeang, Menara UMNO,
Penang, Malaysia

7. M&E SERVICES

The service core provides means of accomodating vertical M&E services runs, such as elevators, duct risers, mechanical pipe risers, hydraulic stacks, electrical and communications cabling, etc (Figs. 11A to 11C). These run either interconnect services in various M&E plant rooms – for example, a gas riser in a low-level gas-meter room to boilers in a thermal plant room at a higher location – or, connect to individual floors from the outset in the case of lifts and air-conditioning ductworks. In many cases, provisions are also made for service connections that may be required by future tenants or occupants.

Air-conditioning shafts

A building's HVAC systems have a major impact on the service core construction. Air conditioning systems in multistorey buildings can be designed as one of two basic types, involving:

- primarily horizontal air distribution from on-floor air-handling units, with vertical pipe risers distributing chilled and heating water
- primarily vertical distribution of cold and hot air in shafts from centrally located air handling units, with the secondary horizontal distribution of air within the false-ceiling void

Vertical air-distribution systems, have some drawbacks especially in relation to their relative inflexibility and difficulties with apportioning costs during out-of-hours operation.

Air-conditioning shaft configuration

As a general principle, air-conditioning shafts approaching the square shape should be more efficient aerodynamically. Whilst this may be true for most supply air shafts, a requirement to fit return air openings with subducts (see 'Protection of penetrations', page 28) may involve some reconfiguration of the shaft to minimise loss of air pressure and therefore allow the designer to select a smaller return air fan and lower the system's energy consumption.

Protection of penetrations

To maintain the integrity of all openings in the service core walls if there is a fire, standard building regulations normally require that all openings in supply air shafts through duct walls are protected by fire dampers, whereas all openings in return air shafts, that also duplicate as smoke-spill shafts, are fitted with specially constructioned ducts – 'subducts' – to prevent the products of combustion re-entering floors unaffected by the fire.

Design of cores for fast-track building construction

In many fast-track projects common in the development of high-rise buildings, head contracts are signed with a builder on the basis of the design consultants' preliminary documentation, which is all that is available at the time. Site works are often started before trade subcontracts with mechanical and other services contractors have been negotiated and put in place, and the builder takes the risk of relying on sketchy documentation that represents only the intent of the design. It is normally the responsibility of the services subcontractor to produce working drawings showing all the dimensions that are critical for construction purposes.

The risks in this approach include lack of feedback from trade subcontractors in relation to adequacy and accuracy of penetrations in core walls. Also, although the consultants' drawings may show a workable solution, it may not be the best one when the speed of construction is assessed.

From the builder's point of view, accessing the inside of the shafts to install services involves additional costs and extra risk. The costs may include scaffolding, or a suspended platform to allow, say, ductwork to be installed. Very often, building regulations also require access openings in shaft walls to ensure an adequate level of fire safety during construction.

For these reasons, alternative approaches aimed at reducing the costs and risks outlined above should be carefully considered. These may include:
- deletion of supply air ducts in supply air risers
- modifications to return air subducts to allow them to be installed from within floors (rather than from inside the shaft)

Temporary services

A further possible constraint on the speed of construction is often associated with temporary building services. These typically involve:
- plumbing to provide water and sanitary plumbing to changing-rooms and canteens
- sprinkler/hydrant pipework to comply with building regulations
- electrical wiring to provide emergency lighting and task lighting
- temporary lift services either using external lifts (alimaks) or utilising the final elevator shaft

By their very nature, these services are a construction consideration and are normally installed by a builder's own plumbers and electricians before services subcontracts are let. Also, temporary services need to be moved during building work to allow the natural progression of construction.

Aspects which need to be considered include:
- location of temporary services in areas that are unlikely to be disturbed as the construction work progresses
- ideally, the use of the final infrastructure and ultimate locations of individual services for temporary purposes, wherever possible

One possibility may include locating temporary services as part of the service core inside stair pressurisation shafts which are normally fitted out very late in the construction process. As the stairwell is usually continuous through the building height in a typical building, this also results in a minimal number of offsets for the services.

Building regulations in some countries do not allow any nonessential services to run through fire-isolated escape routes, including stairwells. This means that temporary installations have to be removed from staircases before a certificate of occupancy can be granted.

Another approach, frequently adopted by builders, is to install temporary services inside high-rise elevator shafts as these will be fitted last by the elevator subcontractor.

Vertical material / people movement

In all high-rise buildings, builders attempt to have temporary lift arrangements available as soon as possible, to reduce reliance on external alimak-type lifts in the basement, or relocated to upper storeys as fit-out of the floors progresses.

Unless the issue is properly addressed during the design stage and adequate provisions made, the resulting restrictions will have a major impact on the speed of floor fit-out.

8. OPPORTUNITIES FOR OPTIMISATION

Significant savings in construction costs can be made by reducing floor-to-floor heights. The task can be accomplished only by addressing the issue on a multidisciplinary basis as this height reduction is normally associated with a potential increase in air-conditioning duct aspect ratio.

The savings are primarily due to the reduction in the direct material costs of the vertical structural elements (mainly, the service core) and the architectural finishes of the facade. However, as an added bonus, the reduced thickness of the service core walls allows the lettable floor space to be increased, thus offering further benefits.

In addition, considerations such as locating vertical air shafts in core corners or in the middle of the core (according to the shape and size of the openings required for duct take-offs to the floors) can reduce costs further.

It is evident from the above that significant cost savings can be achieved when issues normally considered as 'design' are evaluated in conjunction with methods of construction, which primarily concentrate on speed.

Traditional structural optimisation on its own is limiting in terms of potential cost benefits. A major factor in reducing the costs is by the maximisation of the net rentable area, and this can only be achieved through a multidisciplinary approach.

Construction issues need to be assessed at the early stage of design. With regard to the service core, dealing with core-wall thickness transitions has a major impact on construction time and costs. However, other aspects associated with the provision of building services, both temporary and permanent, cannot be ignored. Speed of construction can be greatly improved if a close interdisciplinary link is established between members of the design team, real estate advisers and builders.

9. NET AND GROSS AREA DEFINITIONS

Designing the relationship between the service core and the floor plate in commercial conditions always needs to be done in the context of maximising the net-to-gross efficiency of the building's floor plates. While the exact definitions of net and gross vary from country to country they can generally be considered as follows. The 'net' area of a building is the saleable or rentable area of its floor plates. The 'gross' area is the whole of the floor plate, including the service core. Building-types have different ratios which are expressed as a percentage. An office building would be considered poorly designed if the efficiency was below 75 per cent. A good single floor-plate efficiency in current commercial real estate terms would be between 80 per cent and 85 per cent. This efficiency might even be a fundamental requirement of the design brief and if calculated incorrectly will cost the client considerable revenue.

We should be aware that the overall building efficiency will usually be lower than the floor-plate efficiency because the building has shared facilities which are not reflected on the floors. These are the main entrance lobby, cafeterias, plant rooms and service corridors, car parks and other 'back of house' areas. The ratio of 'net' area to 'total gross area' of the building will provide this efficiency figure.

A useful differentiation is to consider the areas in a typical floor plate for an office building in terms of the following (Fig. 12):

- Gross External Area (GEA)
- Gross Internal Area (GIA)
- Net Internal Area (NIA)
- Net Usable Area (multiple tenancy, single tenancy, double tenancy)
- Plant and maintenance area
- Shell and core area

Gross External Area: The gross external floor area (GEA) in this context refers to the entire floor-plate area measured from the outside edges of the external walls. This figure is useful in defining the entire built-up area of the floor plate.

Net Internal Area: The Net Internal Area (NIA) is the area of the entire floor measured to the inside edges of the external walls.

Net Internal Area (multiple tenancy): This is the internal area measured from the insides of the external walls, or to the inside wall of any internal access passageways, but excludes the service core area.

Net Internal Area (single tenancy): For this the internal area is measured from the inside of the external wall and includes the toilet areas and elevator lobbies in the service core area. This is because some single tenants occupying an entire floor may seek to customise the toilets to their own requirements and to maintain the toilet areas within the occupied floor. This is only feasible if the elevator shafts are fire-pressurised and the elevator openings have fire-rated doors. In this case the elevator lobby does not need to be a fire-protected zone and can become part of the net rentable or usable area.

Net Internal Area (double tenancy): This is the internal area measured from the inside of the external walls, but excludes the elevator lobbies and circulation to the escape staircases.

Plant and maintenance areas: These include the elevator shafts, service riser ducts and plant rooms, areas that are maintained by the building owner or that can be classified as landlord area. In the case of multiple and double tenancy, these areas also include the toilets. This definition is useful in tenancy contracts.

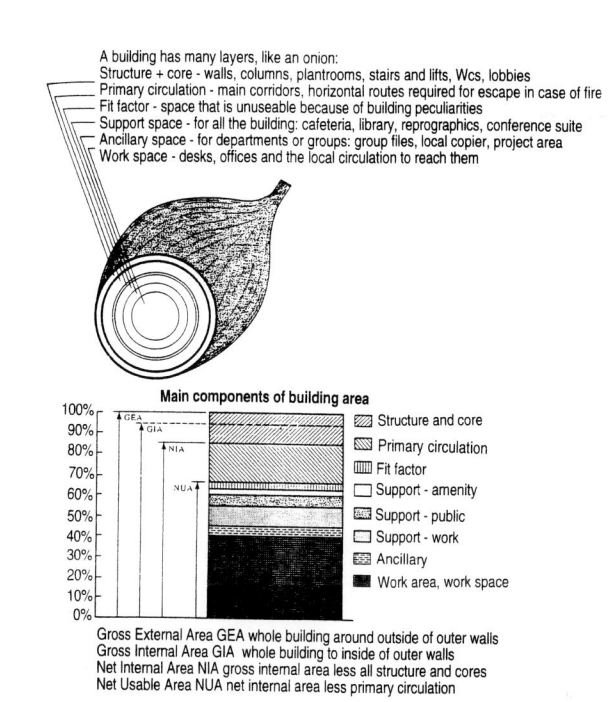

Fig. 12: Floor area measurement

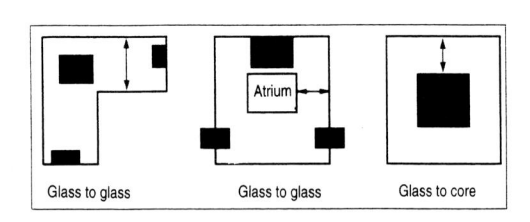

Fig. 13: Building configuration examples

10. FLOOR-PLATE CONFIGURATION

When shaping the floor plate it is useful for the designer to be aware of historical precedents and factors. For instance, until the mid-1960s and 1970s, the classic configuration in most high-rise buildings was to have the service core in the central position. A common floor-plate design was probably about 36 metres square. This would equate to approximately 1300 square metres GEA. With the core in the middle, the distance from the edge of the core to the external wall would be approximately 10.5 metres. The usual level of net-to-gross efficiency achievable per floor-plate would probably be around 80 per cent to 82 per cent (Fig. 13). The usual variation on this configuration in the early years of the high-rise office tower was the rectangular floor-plate with a central service core.

Where the side service core configuration is adopted, the core itself need not be located outside the floor plate but can be recessed in, to be part of the floor-plate shape. This was used by Frank Lloyd Wright in 1915 in one of his skyscraper tower proposals (Fig. 14), and by Louis Sullivan in the Wainwright Building in 1891 (Fig. 15).

Two or more banks of elevators are often used for the side core configuration (Fig. 16). The elevators are designed as a system of double- and sometimes triple-zoned banks with sky lobbies for transfer floors. Once the lower banks of elevators are eliminated after the transfer floor, the space above can be reclaimed for net use, thereby increasing the floor efficiency. Recessing the service core walls within the floor plate further facilitates the reclamation of core areas for net usable areas.

The depth of the floor plate from the external wall to the inside wall of the service core depends on the distance acceptable to the users. A commonly acceptable depth in earlier high-rises might be around 7.5 metres to 10 metres measured from the inside of the external wall to the internal wall of the service core. Generally, high-rise users tend to prefer spaces within an 8 metre to 10 metre zone from the perimeter of the floor plate for reasons of accessibility to windows, sunlight, views and ventilation. This zone is considered to be the most flexible area in the floor plate. Cellular room modules are generally located here with the modules planned and designed around the 1.5 metre grid system in both directions. This ties up with the size of office furniture and with facade systems where partitions can be aligned with window mullions.

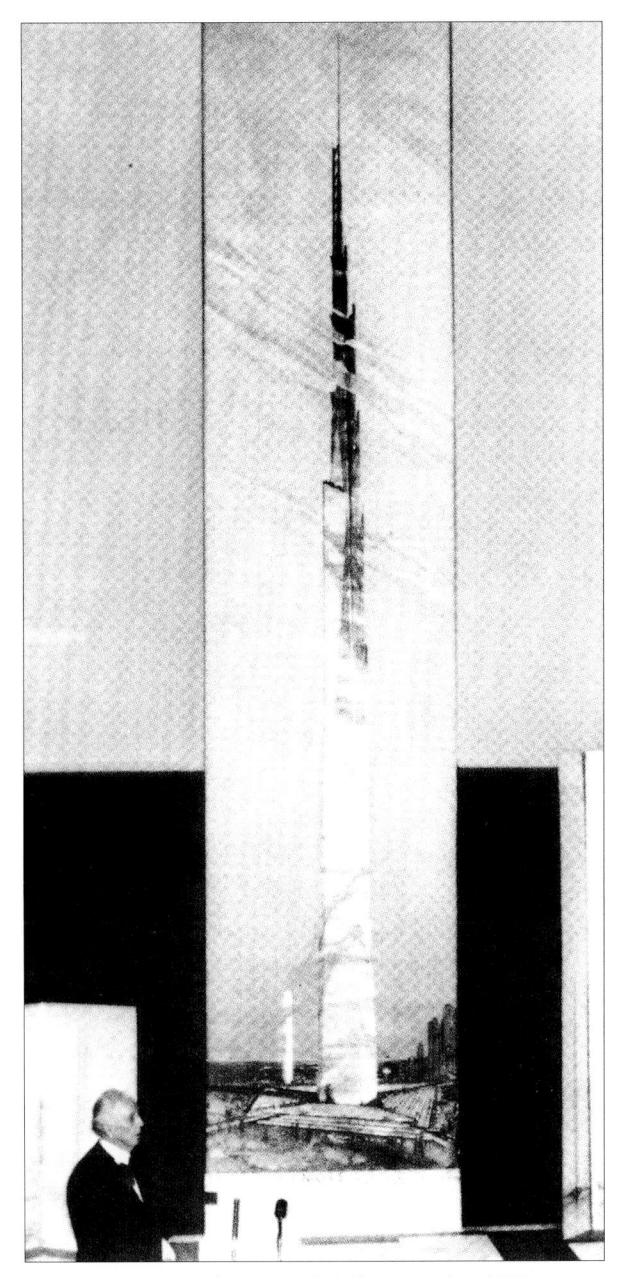

Frank Lloyd Wright, One-Mile-High Tower, project 1956

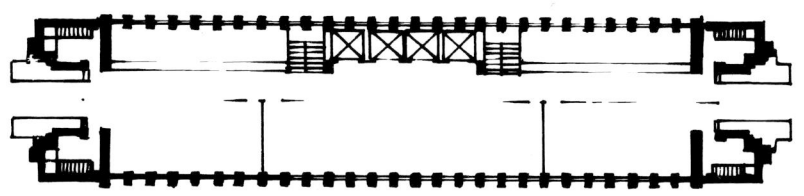

Fig. 14: Recessed side core configuration: Frank Lloyd Wright, Project for Skyscraper, 1915

Fig. 15: Adler & Sullivan, Wainwright Building, St Louis, 1891

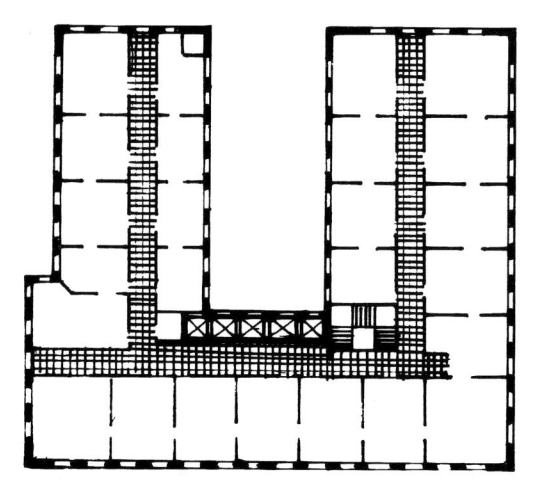

Fig. 16: Recessed core: Adler & Sullivan, Wainwright Building, St Louis, 1891

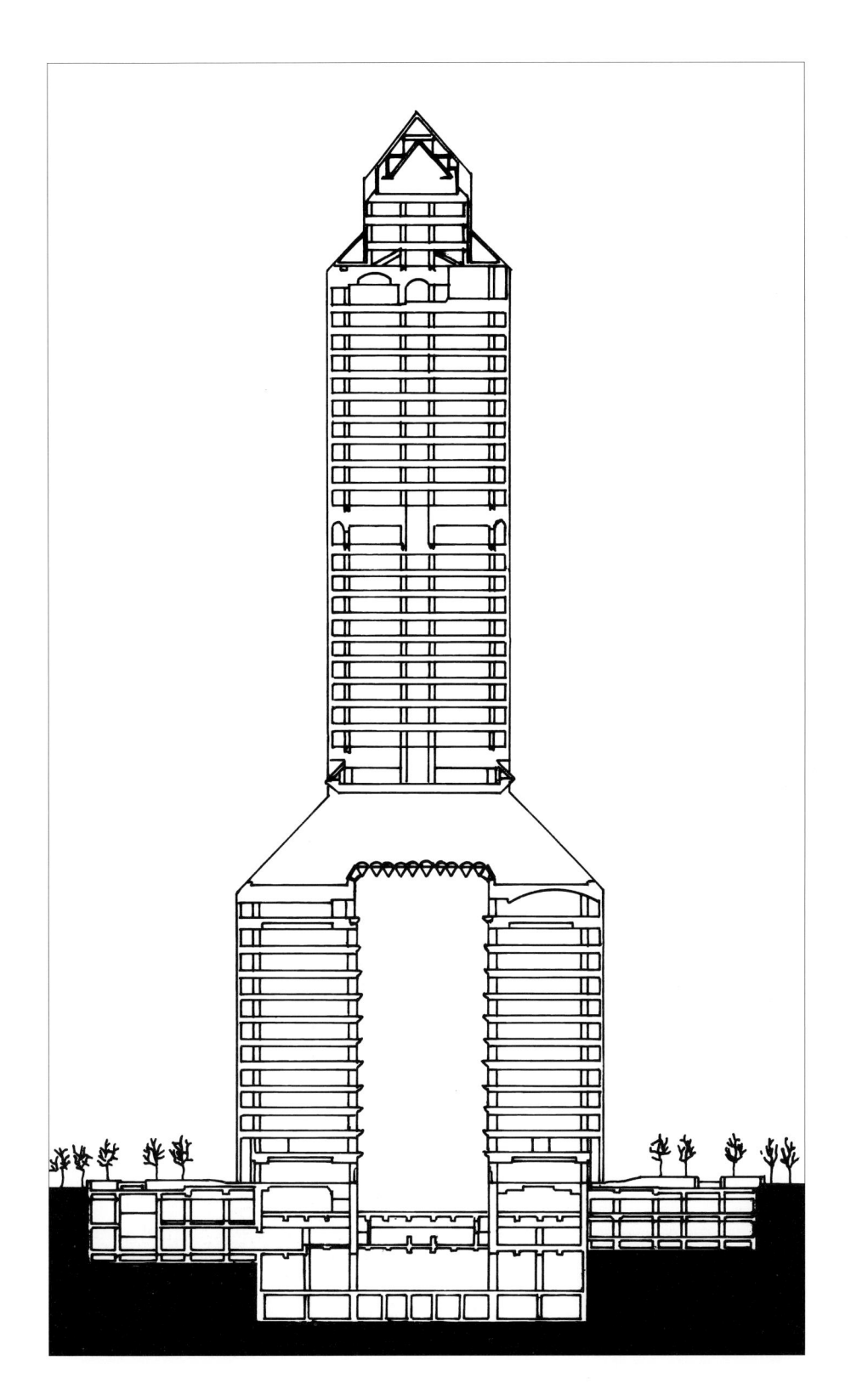

Nikken Sekkei, NEC Tower, Tokyo, 1990

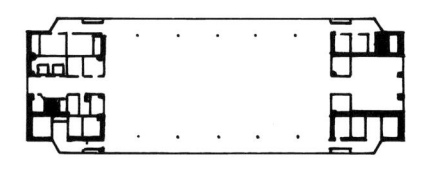

Recent studies in daylighting indicate a trend that the furthest distance from natural daylight for a desk or work surface should be no more than two and a half times the height of the external window. This is approximately 6 metres. Typically this leads to a narrower and more rectangular floor plate configuration.

For example, the Lever House, built in New York in 1952, has a slim rectangular and narrow floor plate with a wall-to-wall depth of approximately 17 metres and an end service core position (Fig. 17). For greater floor plate areas that still have narrow wall-to-wall dimensions the atrium configuration (Fig. 5B) becomes inevitable. There are also local preferences for what constitutes an optimum lettable area (for an office). Generally, this is around 1,200 square metres to 2,000 square metres subject to the tower's built form but this varies from country to country depending on the land economy, the average size of the local central business district or the companies and their spatial provisions.

The floor-plates described above are rectilinear, to demonstrate the subtleties involved in establishing the maximum net usable areas in floor-plate design. Fig. 18B shows that the configuration need not be rectilinear.

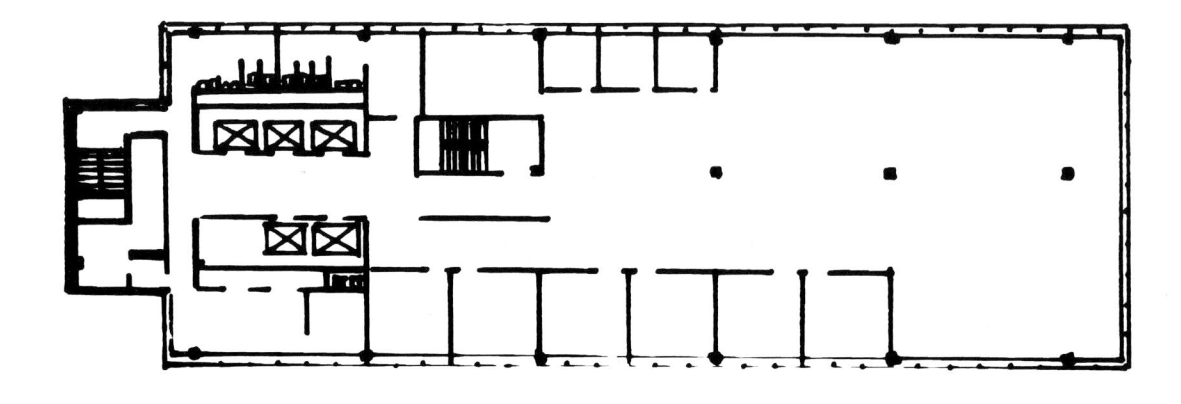

Fig. 17: Lever House, SOM, New York

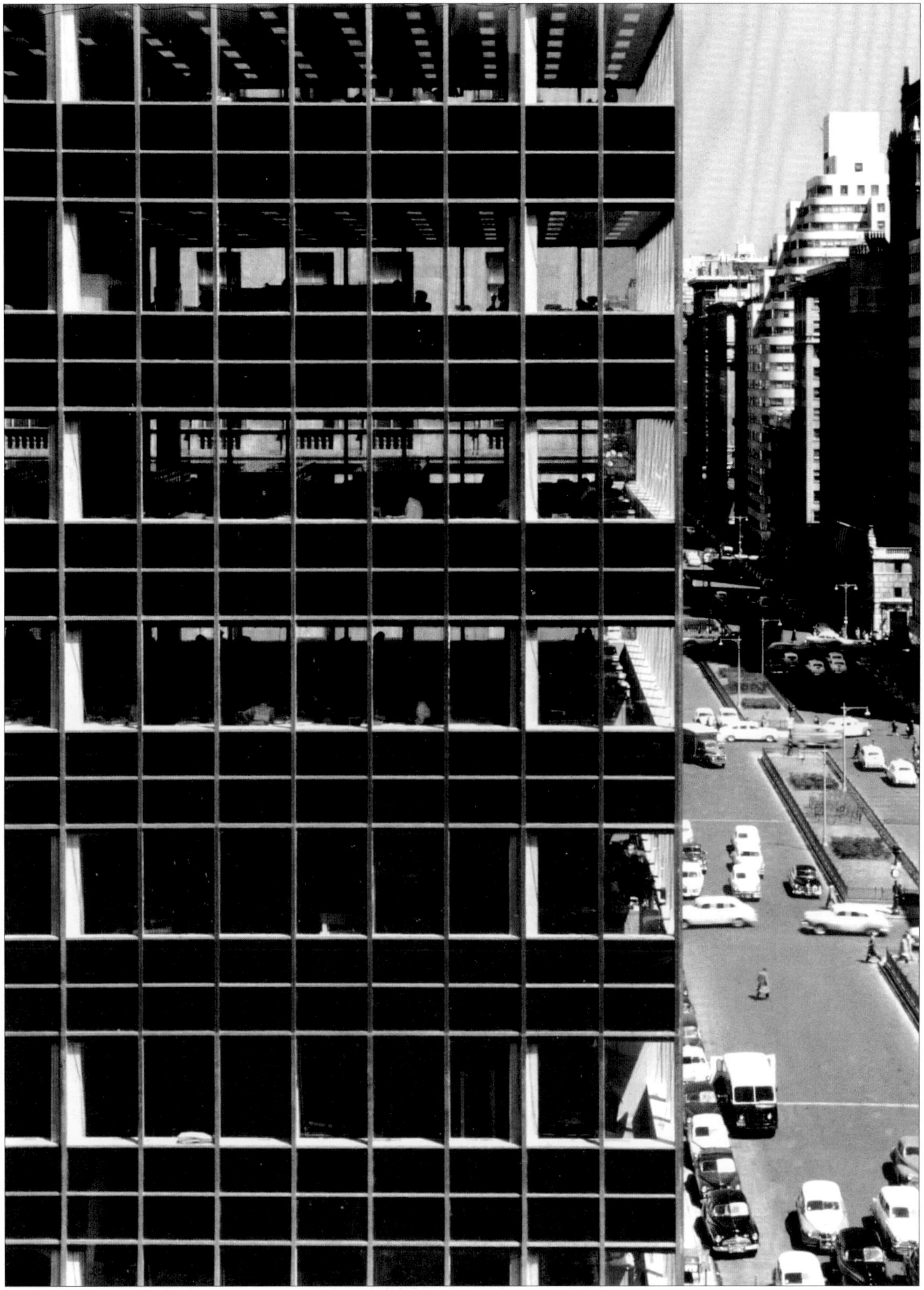

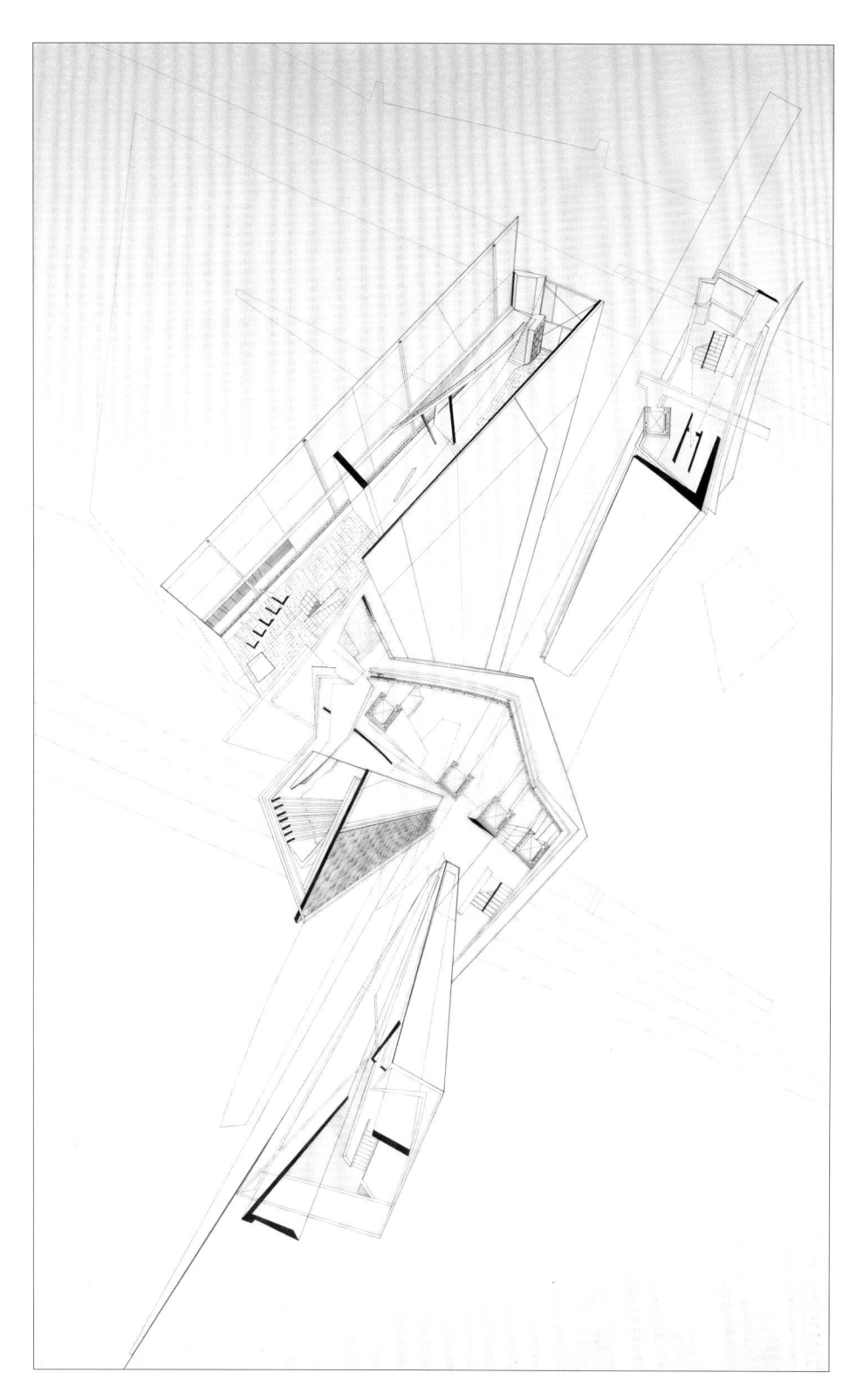

Ben van Berkel, Scuba Towers, birds eye perspective

11. WORKSPACE DESIGN

In today's office buildings the placement of the service core is crucial to the way the internal workspaces are to be organised. Duffy (1997) has identified four major generic organisational types as a shorthand way of capturing the distinctive work patterns and design features characterised as hive, cell, den and club (Fig. 18A). The core placements need to be decided in conjunction with the internal organisational type as an 'inside-out' design activity.

Each type has a different configuration of workstations furniture and levels of enclosure (partitioning). 'Hive' offices can be compared to beehives occupied by busy worker bees; 'cell' spaces resemble those in monks' cloisters; 'dens' are busy and interactive spaces where it is easy to work in teams; 'club' is the new transactional office, similar to a gentleman's club.

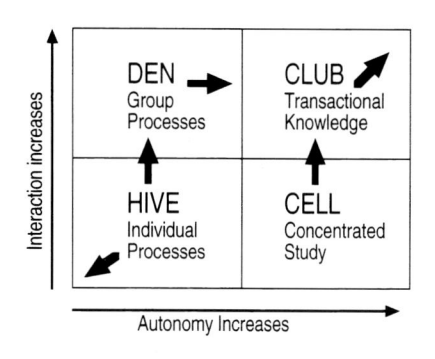

Fig. 18A: Types of workplace configurations for office space planning

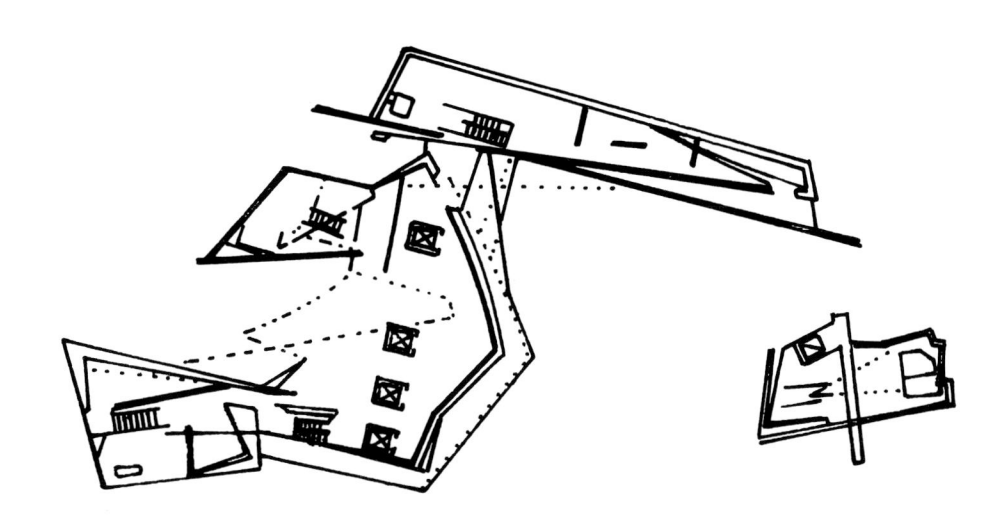

Fig. 18B: Irregular floor plate, Ben van Berkel, Scuba Towers

12. ELEVATOR SHAFT CONFIGURATION

In determining the internal configuration of the service core, one of the first elements to identify is the extent to which vertical transport will be provided within the building. A high-rise building requires a set of elevators and therefore a specialist elevator consultant in the design team. In conjunction with the structural and building services engineers, the architect will look at the elevator grouping and arrangement – including people lifts, goods lifts and fire lifts – that meets design criteria such as average waiting times, handling capacities, etc. These criteria differ depending on the building-type – hotel, apartment block or offices. A large bank of elevators is the main element in a service core design and all other elements are designed around it. Vertical transport solutions are complex, requiring computer simulations of people movements, predictions about users within buildings and historical data from existing buildings. The subject is specialist enough to deserve coverage outside the scope of this book.

The vertical transportation of people within a high-rise building will also depend on local fire regulations. The fire department may require fire compartmentation between the elevator lobby and elevator shafts. A separate fire-fighting elevator capable of moving fire-fighters around a burning building when all other lifts have returned to their neutral position, is often required. The relationship between these elements may affect the lettable area of the building, hence the need for the design team to find the optimum solution while not restricting themselves to traditional and conventional elevator shaft configurations. Figs. 19A to 19E show possible alternatives.

While these are, again, conventional rectilinear layouts, they are simply illustrative of planning principles and should not preclude other more experimental options (Fig. 20). It is also important to remember that fire-protection considerations must be taken into account in the organisation of elevators and escape stairs. Requirements vary depending on local building regulations.

Two-car Groupings

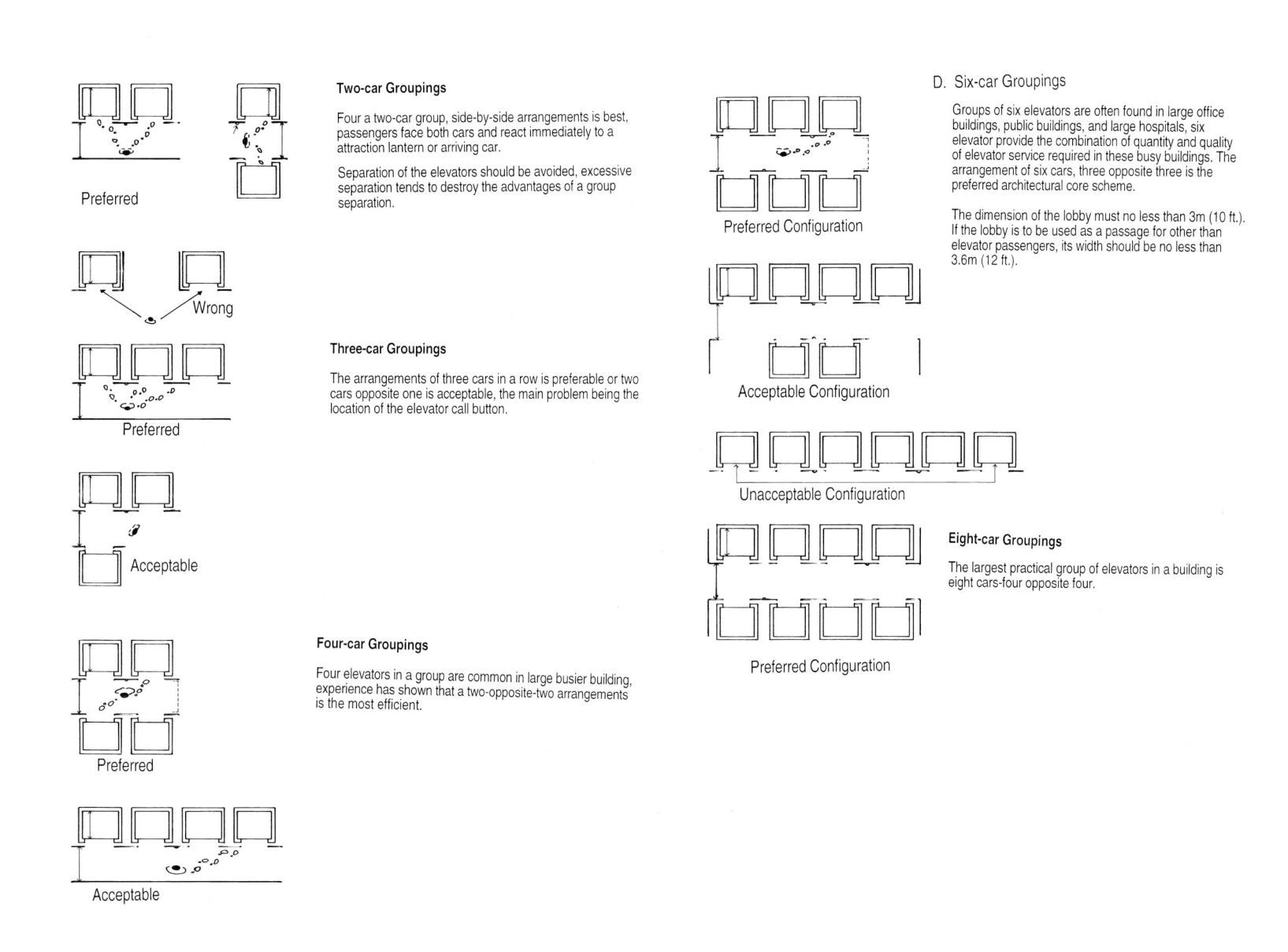

Four a two-car group, side-by-side arrangements is best, passengers face both cars and react immediately to a attraction lantern or arriving car.

Separation of the elevators should be avoided, excessive separation tends to destroy the advantages of a group separation.

Three-car Groupings

The arrangements of three cars in a row is preferable or two cars opposite one is acceptable, the main problem being the location of the elevator call button.

Four-car Groupings

Four elevators in a group are common in large busier building, experience has shown that a two-opposite-two arrangements is the most efficient.

D. Six-car Groupings

Groups of six elevators are often found in large office buildings, public buildings, and large hospitals, six elevator provide the combination of quantity and quality of elevator service required in these busy buildings. The arrangement of six cars, three opposite three is the preferred architectural core scheme.

The dimension of the lobby must no less than 3m (10 ft.). If the lobby is to be used as a passage for other than elevator passengers, its width should be no less than 3.6m (12 ft.).

Eight-car Groupings

The largest practical group of elevators in a building is eight cars-four opposite four.

Figs. 19A to 19E: Elevator car configuration options

43

	Burolandschaft offices	Traditional British speculative offices	New 'Broadgate' type of British speculative office	Traditional North American speculative office	The new North European office
No. of storeys	5	10	10	80	5
Typical floor size	2,000 sq.m.	1,000 sq.m.	3,000 sq.m.	3,000 sq.m.	Multiple of 2,000 sq.m.
Typical office depth	40m	13.5m	18m and 12m	18m	10m
Furthest distance from perimeter aspect	20m	7m	9-12m	18m	5m
Efficiency: net to gross		80º	85º	90º	70º (lots of public circulation)
Maximum cellularization (% of useable)	20º	70º	40º	20º	80º
Type of core	Semi-dispersed	Semi-dispersed	Concentrated: extremely compact	Concentrated: extremely compact	Dispersed: stairs more prominent than lifts
Type of HVAC services	Centralised	Minimal	Floor by floor	Centralised	Decentralised minimal use of HVAC

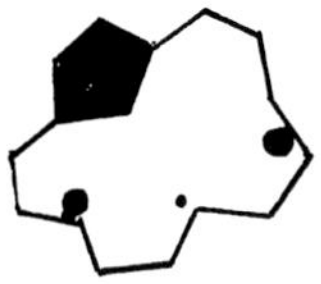

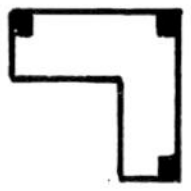

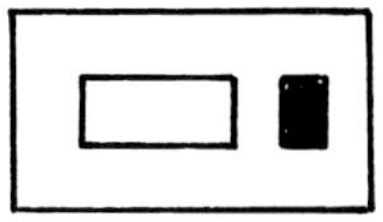

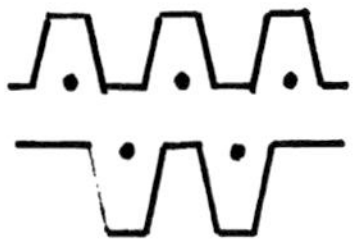

Fig. 20

13. ELEVATOR SHAFTS WITHIN THE SERVICE CORE

Once the location of the service core on the floor-plate has been determined, the exact size of the core (internal shaft dimensions, wall thickness, etc.) needs to be established to calculate the area efficiencies. It is next necessary to define to which the services duct and shaft sizes, and more importantly the elevator system, should conform. Early liaison with the fire officer is important in establishing life-safety requirements. Elevator shaft dimensions can easily be obtained for all the components of the elevator system. Common fire compartmentation of all vertical shafts can minimise wall thickness provided the structural designer is satisfied with the core stability in the overall design of the building. Elevator shafts are sized according to car shapes and sizes and door sizes, with due consideration given to space requirements for guide rails and brackets, counterweight systems, running clearances and ancillary equipment. Sufficient air space should always be provided around cars and elevator counterweights to minimise buffeting and airborne noise during operation.

In organising the configuration of elevator banks in the service core, it is necessary to ensure that a bank of two, three or four elevators in line shares a common fire-protected shaft without a dividing structure, so avoiding a single enclosed elevator shaft. If single enclosed elevator wells are necessary for structural reasons, the designer must ensure air relief slots (ideally, full height vertical slots) to allow adequate air relief.

'Outward facing' elevators (elevator-bank openings that face directly into the net usable area) are the most efficient (Fig. 4). This is because the lobby is accountable as part of the net usable area on most typical floors. However, in certain countries local building codes permit this layout only if the service core has fire-rated elevator doors and pressurised elevator shafts. Such an arrangement also allows for good access, with wide elevator lobbies at ground-floor level to handle traffic peaks more efficiently.

The lobby of 'inward facing' elevators (or two banks of elevators facing each other) can be included as part of the core but the arrangement is less efficient in terms of net usable area versus gross floor area. For inward facing elevators, the designer must ensure both ends of the lobby are kept open.

As a general guide, the width of the elevator lobby should be twice the depth of the elevator cars it is servicing. For a single line of elevators, a minimum lobby width of

2.5 metres is generally recommended. When designing the service core in relation to the floor-plate, the designer must ensure that the lobby will not be used as a common or public thoroughfare at ground-floor level.

In multi-use buildings care should be taken to provide separately identified lobbies for each group of elevators, particularly on the ground floor where clear signage is essential. This is also applicable to high-rise buildings where the lower floors are served from a separate bank of elevator (Fig. 21).

Single Zone — Single Bank — Double Bank

Two Zone — Common Lobby — Separate Lobby

Multiple Zone — Zone 1 — Zone 2 — Zone 3

Multiple Zone/Sky Lobby — Express — Zone 1 — Zone 2 — Zone 3 — Express

Fig. 21

14. STAIRCASES, EXITS AND LIFE-SAFETY CONSIDERATIONS

The other vertical circulation component and key element in designing the service core is the staircases. Their location, as a required means of egress, is often one of the decisive form-givers in any major building. The escape stairs are separate from the ceremonial internal staircases.

For skyscrapers, elevators are not considered 'legal exits' in a fire emergency because they may fail to operate and can become lethal devices by delivering people to the wrong floor or, worse, trapping them. Moreover, the doors of elevator shafts cannot close tightly and the shafts, often filled with smoke, can connect the floor containing the fire to the rest of the building unless they are fire-pressurised and/or have high-fire-rated doors). Usually the fireman at the fire-control command room immediately brings all elevators down to the ground floor and uses the designated firemen's elevator for fire-fighting. It is the building's staircases that are the critical parts of its life-safety system. It follows that both their number and location are crucial in the design of the service core.

Although local building regulations go into great detail about the exit requirements and the way in which exiting enclosures must be constructed, the following key points have a major impact on floor-plate configuration and the service core design:
- use of the building (office, apartment block, store, hospital, etc)
- total number of people in the building (determines the number of separate exits)
- the provision of fire-escape exits that lead occupants to a safe area
- limitations on the maximum travel distance permitted to reach a fire-protected exit enclosure
- the provision of a choice of paths to an exit, and a choice of exits in case one is blocked

The first three items are usually taken directly from local building codes. The last two items require proper proportioning and shaping of the internal floor-plate configuration in order to comply with a specific maximum 'dead-end corridor' length. The proportioning can have a dramatic effect on the shape of the floor plate and hence the overall shape of the building. For example, as most building regulations do not permit any

dead-end corridors in a hospital the stairs are usually located at the ends of the building. Some fire authorities adopt the following premises for life-safety systems:
- after about three minutes in a smoky situation humans will faint or be asphyxiated
- humans can generally move about 12 metres per minute during a fire with thick smoke
- in certain areas involving 'deadends', occupants must be evacuated within one minute and the length of the deadend corridor must not be more than 12 metres

Depending on the distance between the furthest point of the floor plate and the staircase, there should be at least two escape stairs in a building so that if one of the staircases is unusable – catches fire for example – the occupants will be able to escape by the other. The stairs must be fire-protected, or fire-compartmented in relation to the rest of the building, and act as safe havens. The building must be designed in such a way that:
- the fire-protected stairs have a stair lobby and smoke lobby. The smoke lobby acts as an initial protection before the heat and smoke penetrate the escape-stairs enclosure.
- smoke from a fire does not enter the escape-stairs lobby as it can spread to the whole building and firemen will be unable to enter the building.
- the escape-stairs lobby and smoke lobby are placed where firemen will put on their breathing apparatus and prepare to enter the building by the stairs involved.
- the escape-stairs and fire-protected areas have the relevant fire-rated doors – half hour, one hour, two hour or four hour for example. The doors prevent smoke and heat penetrating the escape-stairs enclosure, so that there is a clear passage for escape, and also to enable firemen to conduct internal fire-fighting effectively.

15. TOILETS IN SERVICE CORES

The toilets are very often located within the service cores (Fig. 22) for ease of plumbing and accessibility to the vertical risers. The extent and number of male, female, executive and disabled facilities are calculated, following local building codes, according to the net area of the floor plate and its projected level of occupancy, and the owner's preferences for any supplementary provisions beyond the usual requirements of the building regulations. Owners usually provide pantries and cleaners' cupboards within this area.

In the event of single occupancy of the floor plate entry to the toilets might be organised so that users are able to access them without going through the elevator lobby. In some buildings, if the toilets are directly accessible from single tenancy spaces the owners often negotiate for these to become part of the net rentable areas.

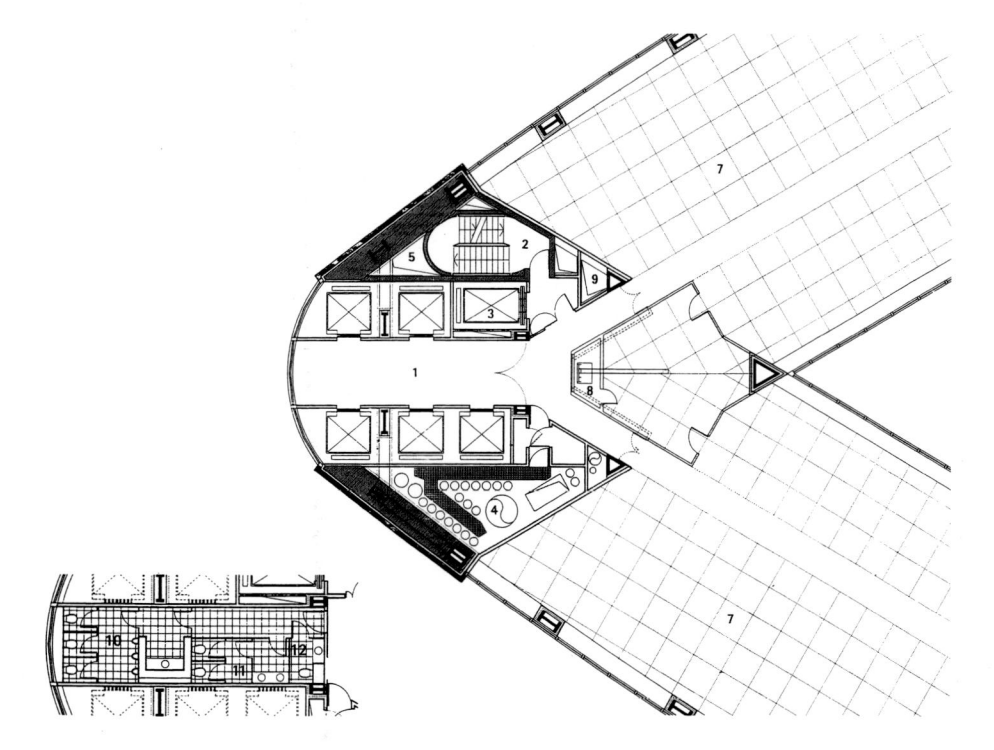

Fig. 22: Toilet layout example: Foster and Partners, Commerzbank, Frankfurt
1. Lift lobby
2. Escape stair
3. Fire lift shaft
4. Main services riser
5. Stair pressurisation duct
7. Office area
8. Document hoist
9. Wet riser
10. Male WC
11. Female WC
12. Disabled WC

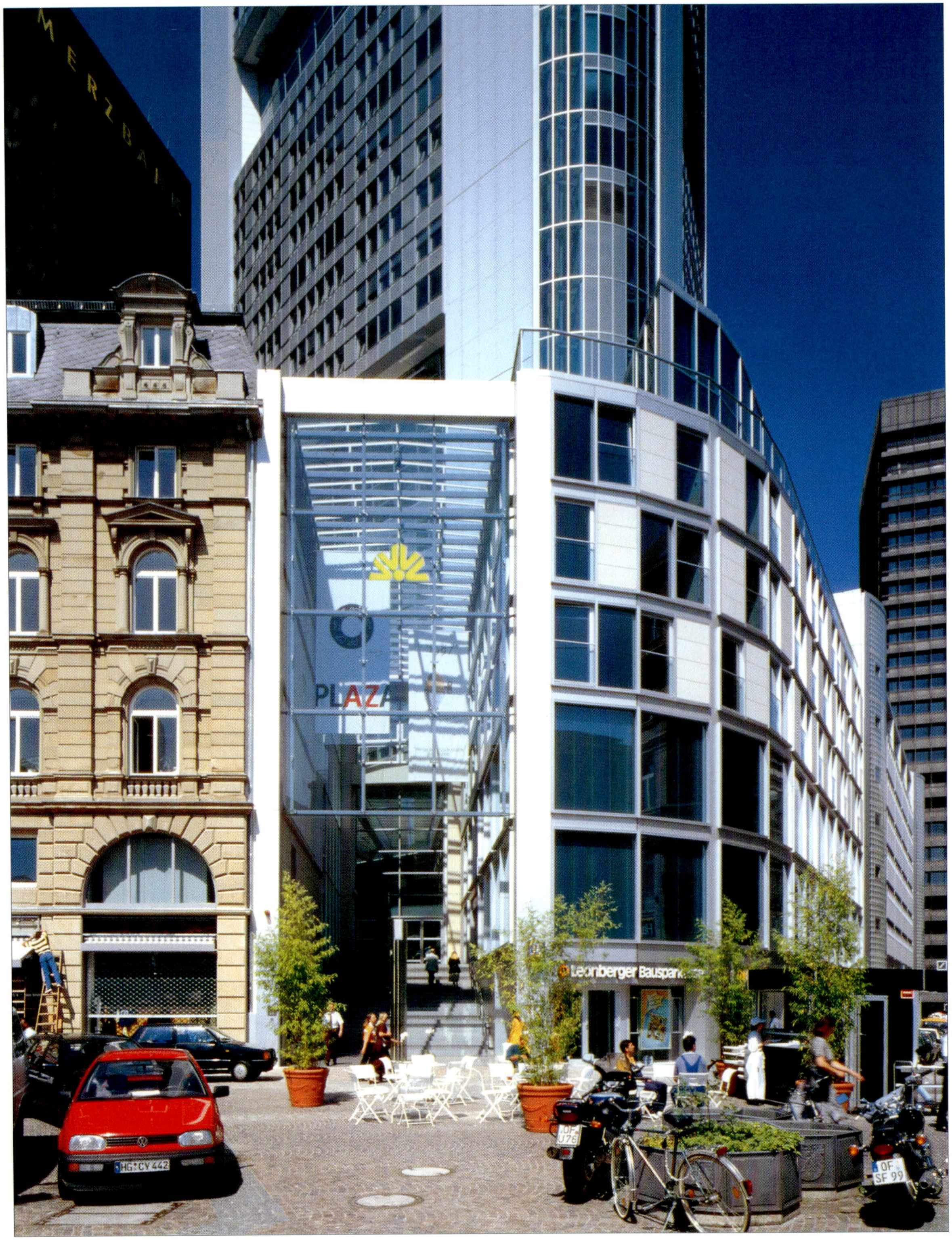
PLAZA

Foster and Partners, Commerzbank, Frankfurt

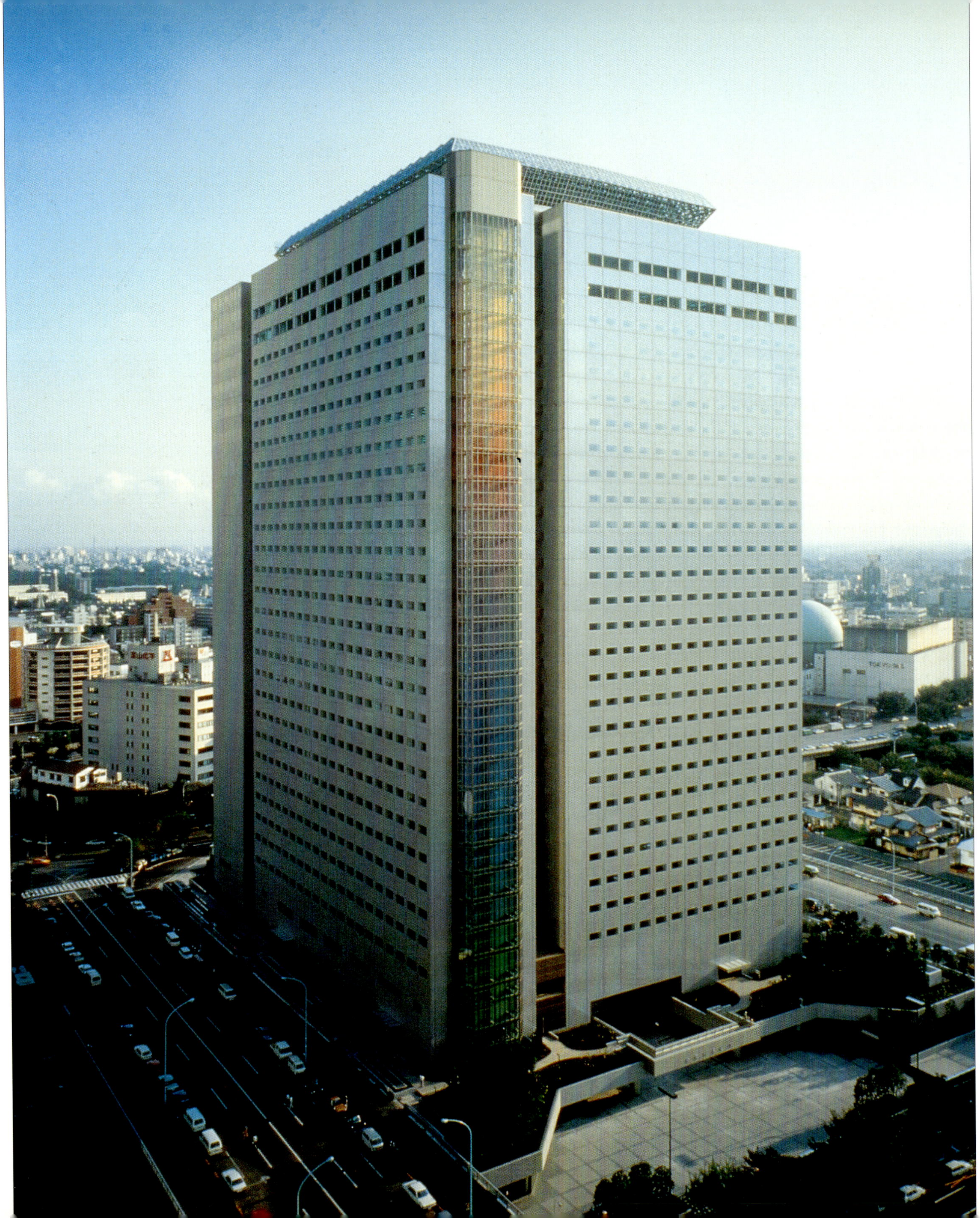

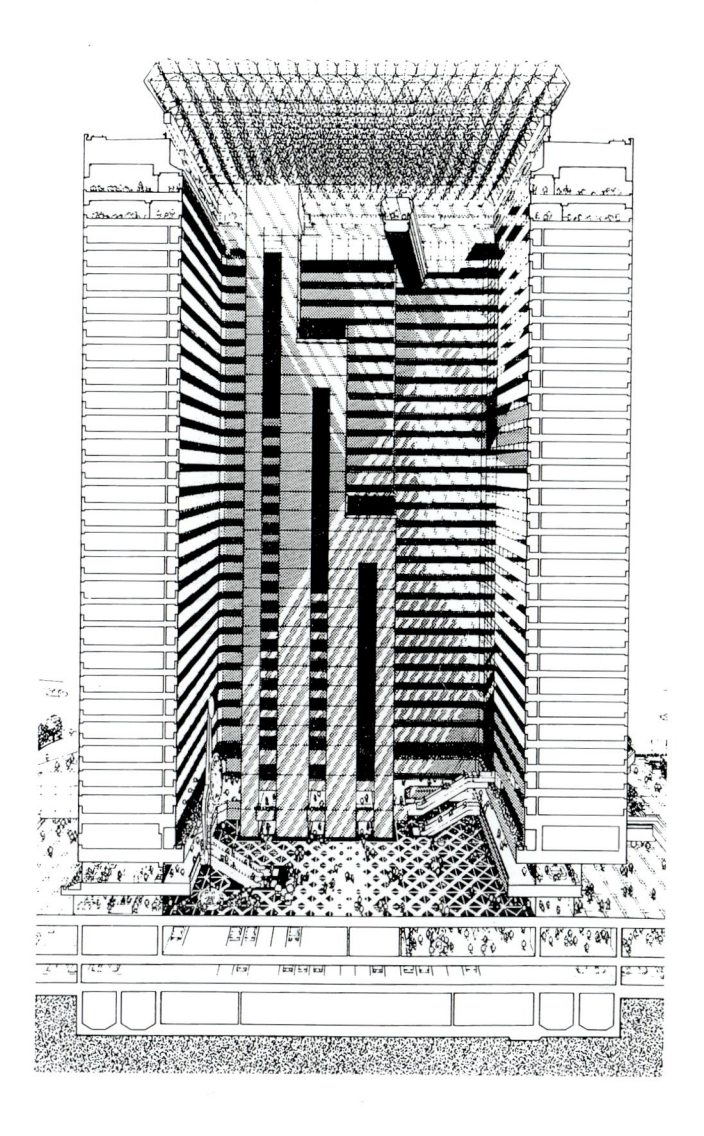

NIKKEN SEKKEI, Shinjuku 'NS' Building, Tokyo

The Shinjuku 'NS' Building, one of the high-rise buildings located in a redevelopment area in Shinjuku subcentre, has been planned to have a court in its centre with two L-shaped office blocks. The court has natural lighting through the glass roof at a height of 130 metres, and provides a new type of common space. The plan of the office blocks is such that the corridors and elevators face the court while all the offices are laid out towards the outside. The building is clad with mirror glass and special tiles with high reflection efficiency. Three underground floors are occupied by rooms for mechanical systems, parking space and an exhibition space. Above the ground the 1st and 2nd floors are for shops, the 3rd to 28th are rental space for offices and the and top 29th and 30th floors are for restaurants. A hanging corridor crosses over the court on the 29th floor.

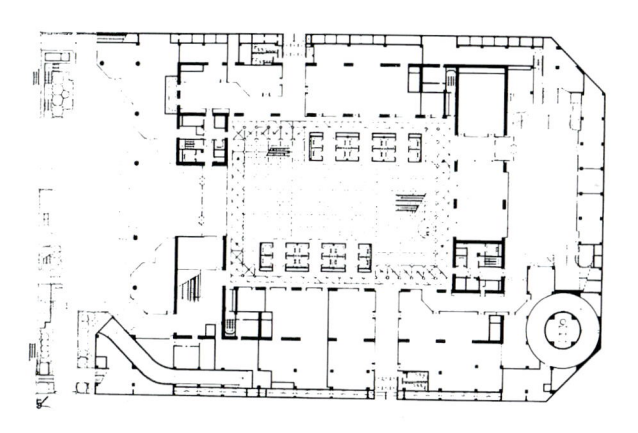

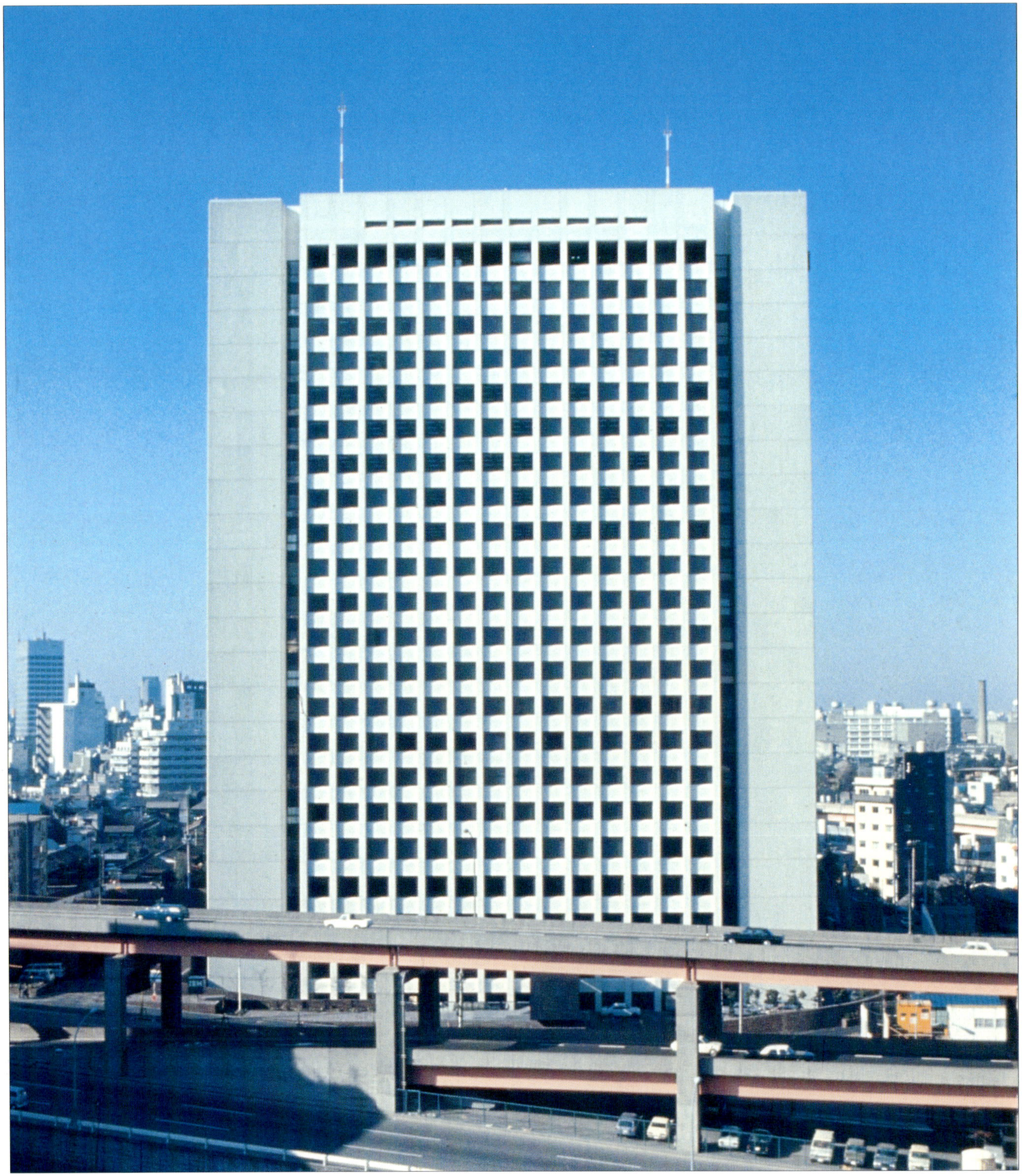

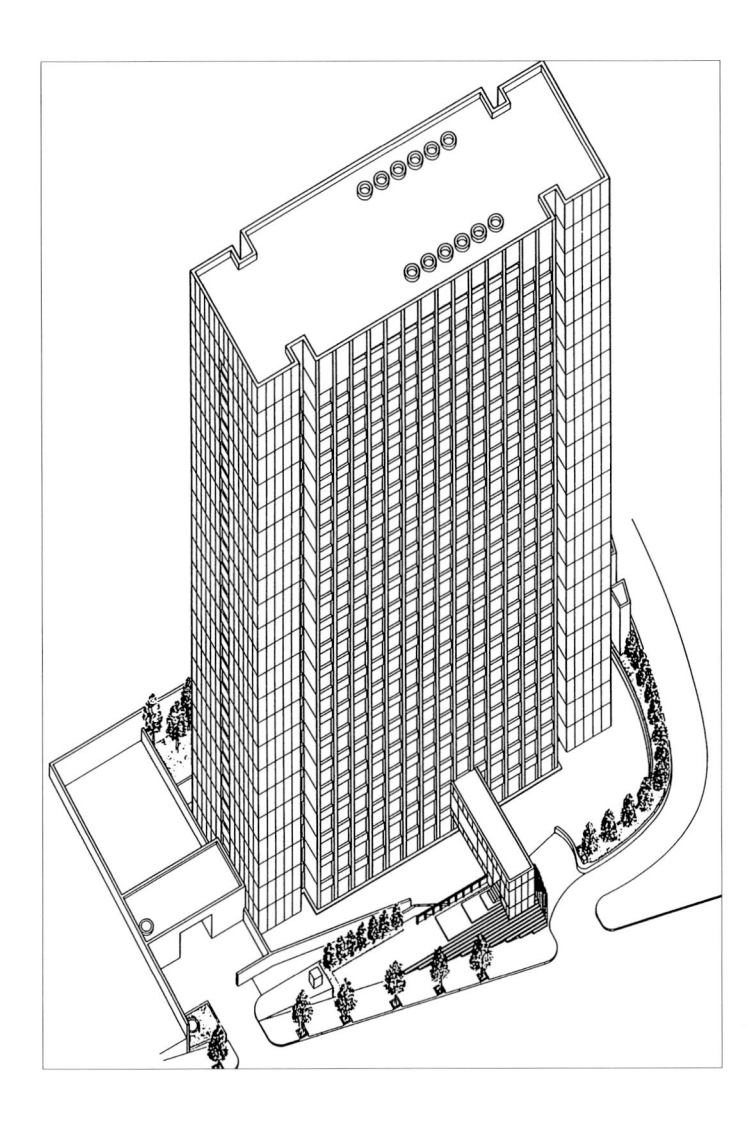

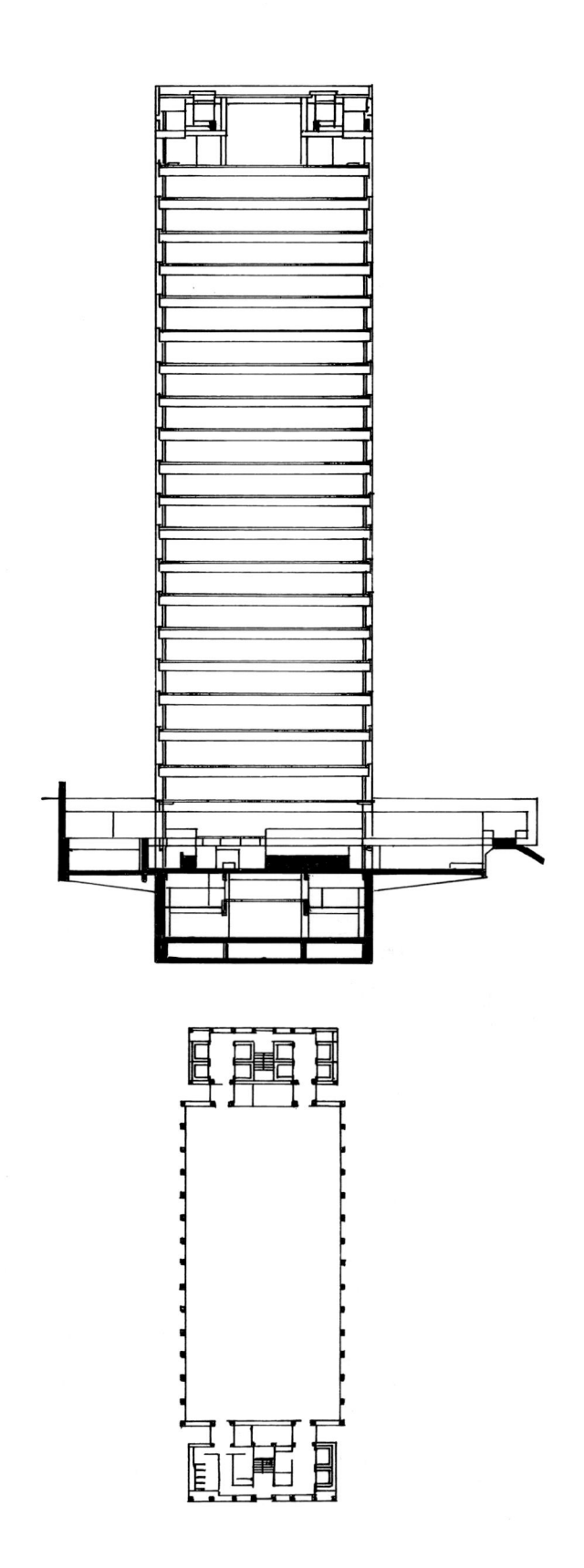

NIKKEN SEKKEI, IBM HEADQUARTERS BUILDING, Tokyo

The new headquarters for IBM Japan stands at the perimeter of the government office area of Kasumigaseki. The office tower in the centre of the plan and the service cores flanking it are structurally independent. All the floors are column-free to allow for maximum flexibility of office layout. Each floor has a floor area of 1000 square metres and a 23 square metre span. The building is based on a 3 square metre module. The interior is uniform oyster grey and the main facade consists of white precast-concrete panels. A heat-recovery system is employed to use energy efficiently and reduce environmental pollution.

NIKKEN SEKKEI, Shinjuku Sumitomo Building, Tokyo

Shinjuku is one of the most important subcentres in Tokyo. Every day over two million passengers pass through the several train stations located there. In front of the stations is a large redevelopment area consisting of 32 hectares of land for 11 high-rise buildings and spacious parks and gardens. The Shinjuku Sumitomo Building, one of the high-rise buildings, uses a striking truncated-triangular plan and double-tube construction system. In addition, it employs a newly developed floor-panel system in which the floors and beams are united. This system made it possible to reduce both the weight of the structure and the time it took to construct it. In accordance with regulations established by all participants in the redevelopment project, public parking space has been provided in the basement, and approximately half the ground level the site has been left open. The exterior of the building is covered with an aluminium curtain wall.

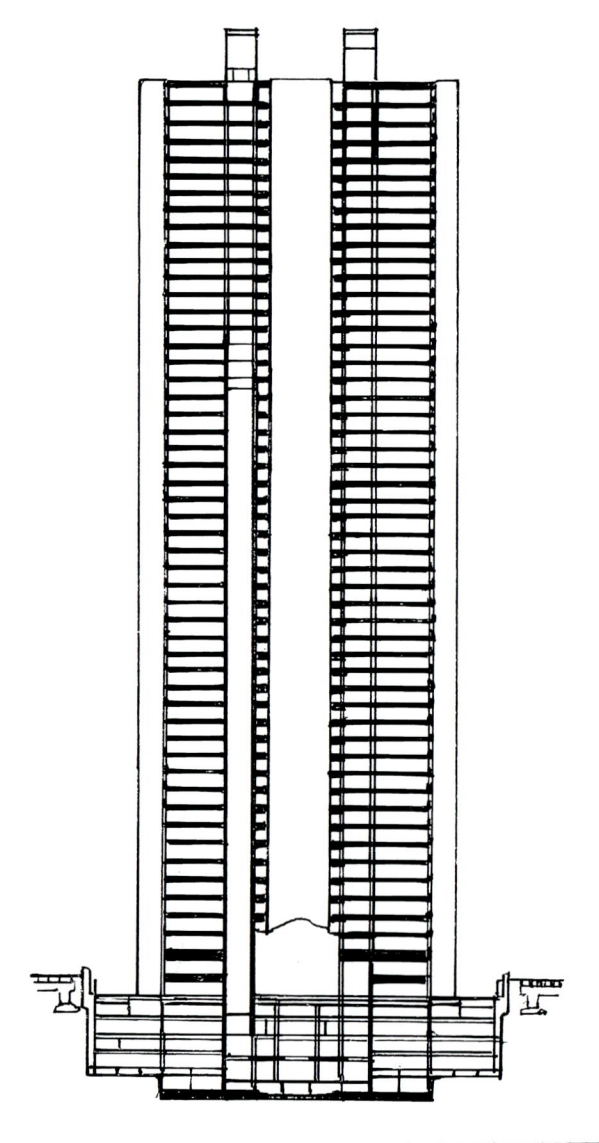

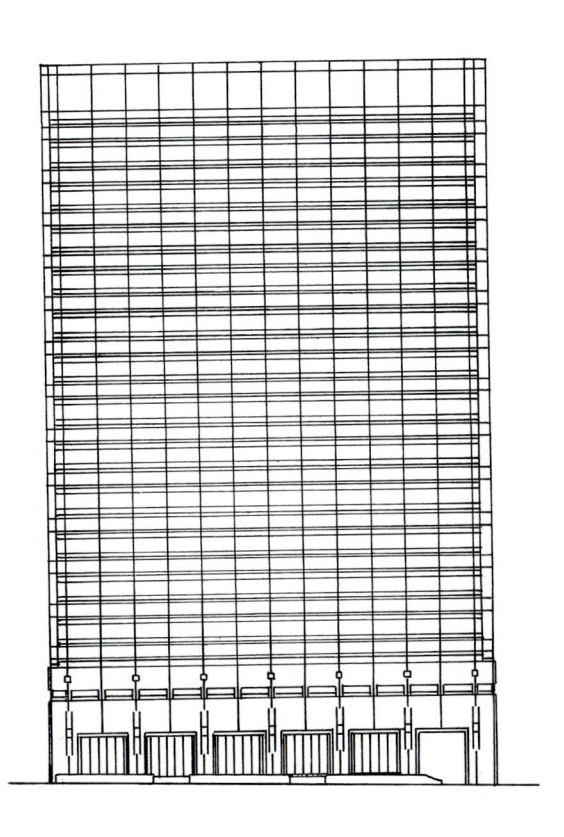

NIKKEN SEKKEI, MITSUI MARINE & FIRE INSURANCE, Nagoya

This office building is located in the business centre of Nagoya. The client's requirement was to realise an impressive building with dignity suitable for use as their head office. The office environment makes good use of advanced intelligence systems. An aluminum curtain-wall construction combined with granite stone and mirrored glass has been adopted in the building's exterior, while using granite-cast pre-cast concrete panel for the lower base.

Two key characteristics of the Mitsui Marine and Fire Insurance building are that it has structured parking systems with the capacity of 150 cars (in this area it is exceptional for such a scale of building as this one) and its service core is laid out according to the shape of the site.

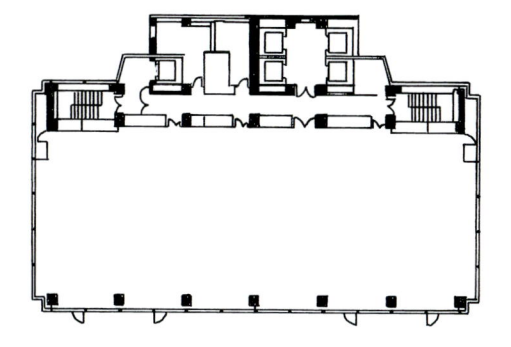

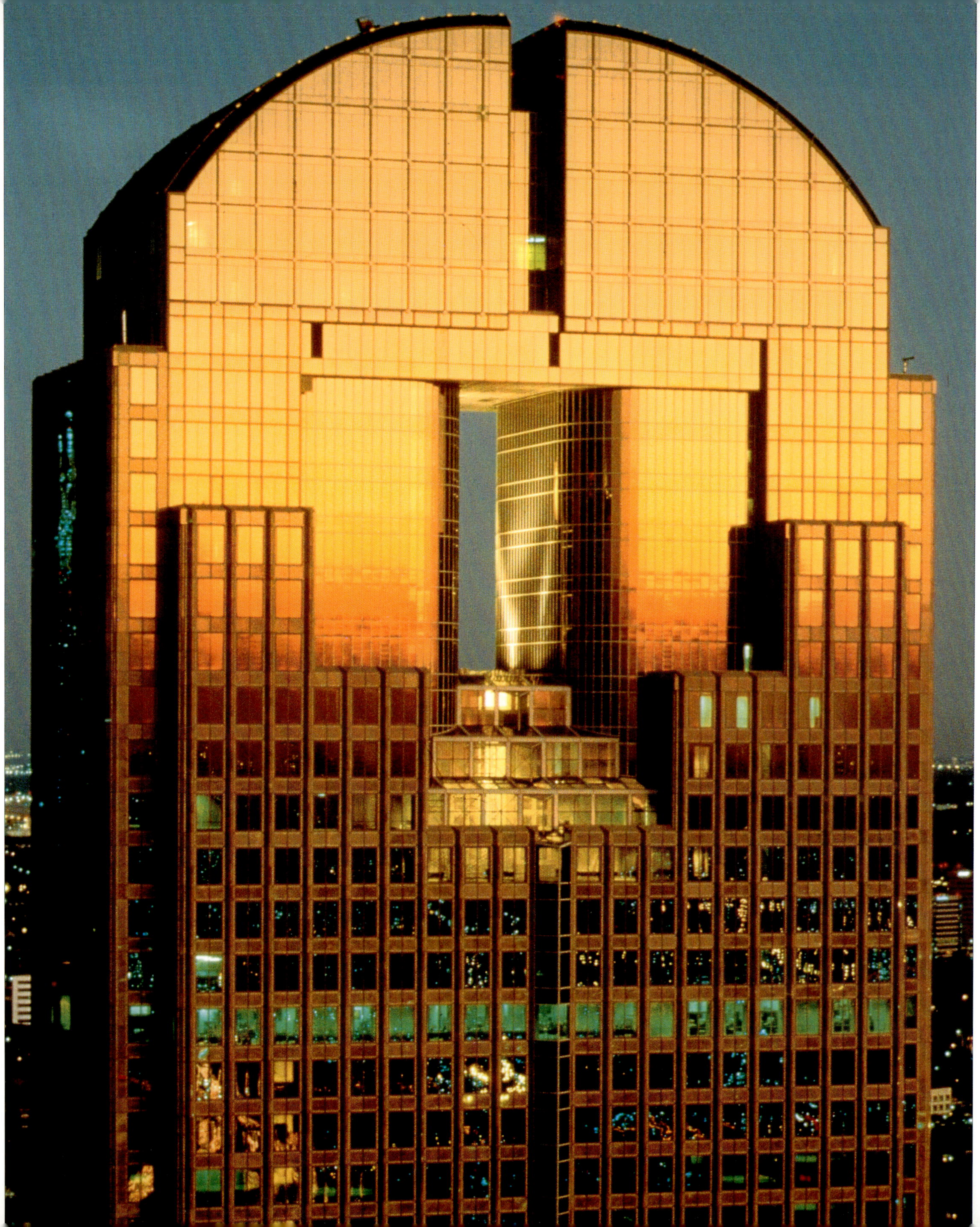

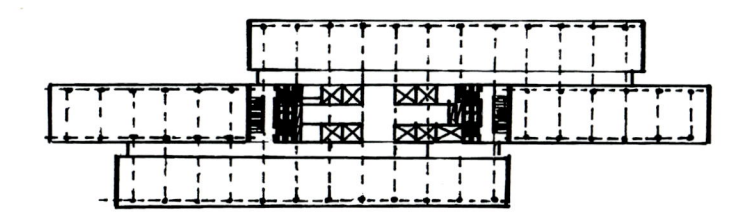

SOM, TEXAS COMMERCE TOWER, Dallas, Texas

Texas Commerce Bank, Dallas, occupies 100,000 square feet of the lower levels, ground floor and third through sixth floors of the Texas Commerce Tower at 2200 Ross Avenue in the main business district. The banking hall was designed to accommodate the client's functional needs as well as to meet their desire for a traditional design evocative of the grand banking hall of TCB's Houston headquarters. The bank's leasing commitment prior to the commencement of construction on the building allowed the interior architects to redesign the lobby spaces with the client's needs and preferences in mind. Major changes to the basic building design included reconfiguring the mezzanine to its current elliptical shape, relocating the escalators which serve the mezzanine, cladding the core walls in limestone and marble and changes to the floor paving and ceiling treatment.

SOM, LEVER HOUSE, New York

Commissioned as a corporate headquarters, Gordon Bunshaft's innovative design established a new vocabulary, and set standards, for office design around the world. Its completion sparked the transformation of Park Avenue between Grand Central and 59th Street from a residential area of masonry buildings to a commercial avenue of glass-and-steel structures.

The 14-storey, 275,000 square foot glass-and-steel tower is set perpendicular to Park Avenue and rests on a two-storey podium. This occupies the entire site and consists of a ground-floor plaza and entrance lobby and second-floor roof garden.

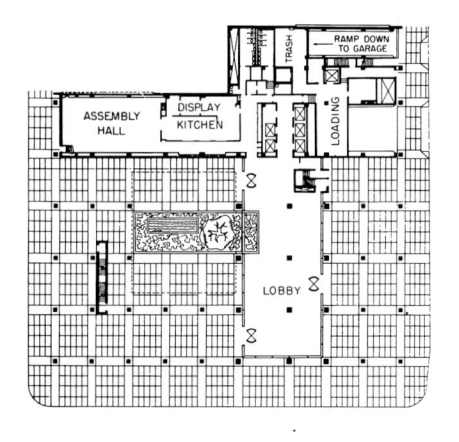

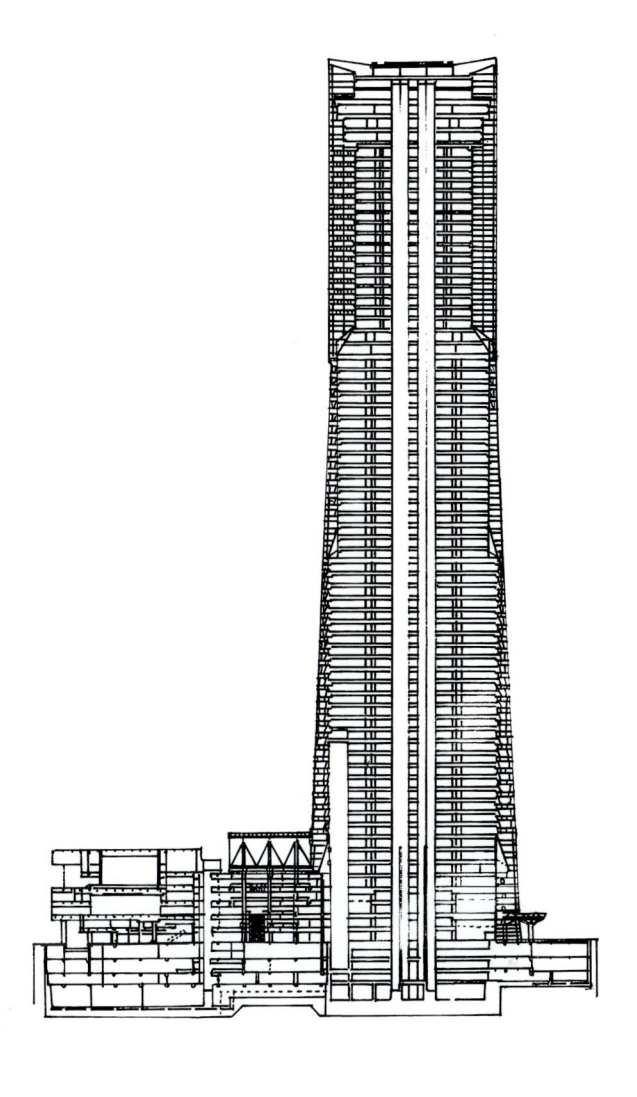

MITSUBISHI ESTATE, THE LANDMARK TOWER, Yokohama

The owner of this tower intended to make a valuable contribution not only to the present generation of inhabitors but to societies many generations into the new millennium. Despite its scale, the designers have attempted to ensure that the spaces within are of human scale. There is an extensive range of functions and facilities including office space, a hotel, a shopping mall and an events hall, all of which are spread across the building's 70 floors above ground and four floors below. The total floor space of nearly 400,000 square metres was constructed from steel frame and reinforced concrete between early 1990 and mid-1993. It still has the appearance of a completely contemporary building and has proved to be economically successful.

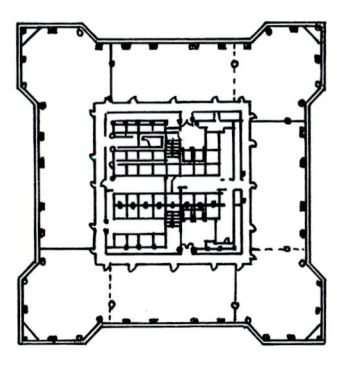

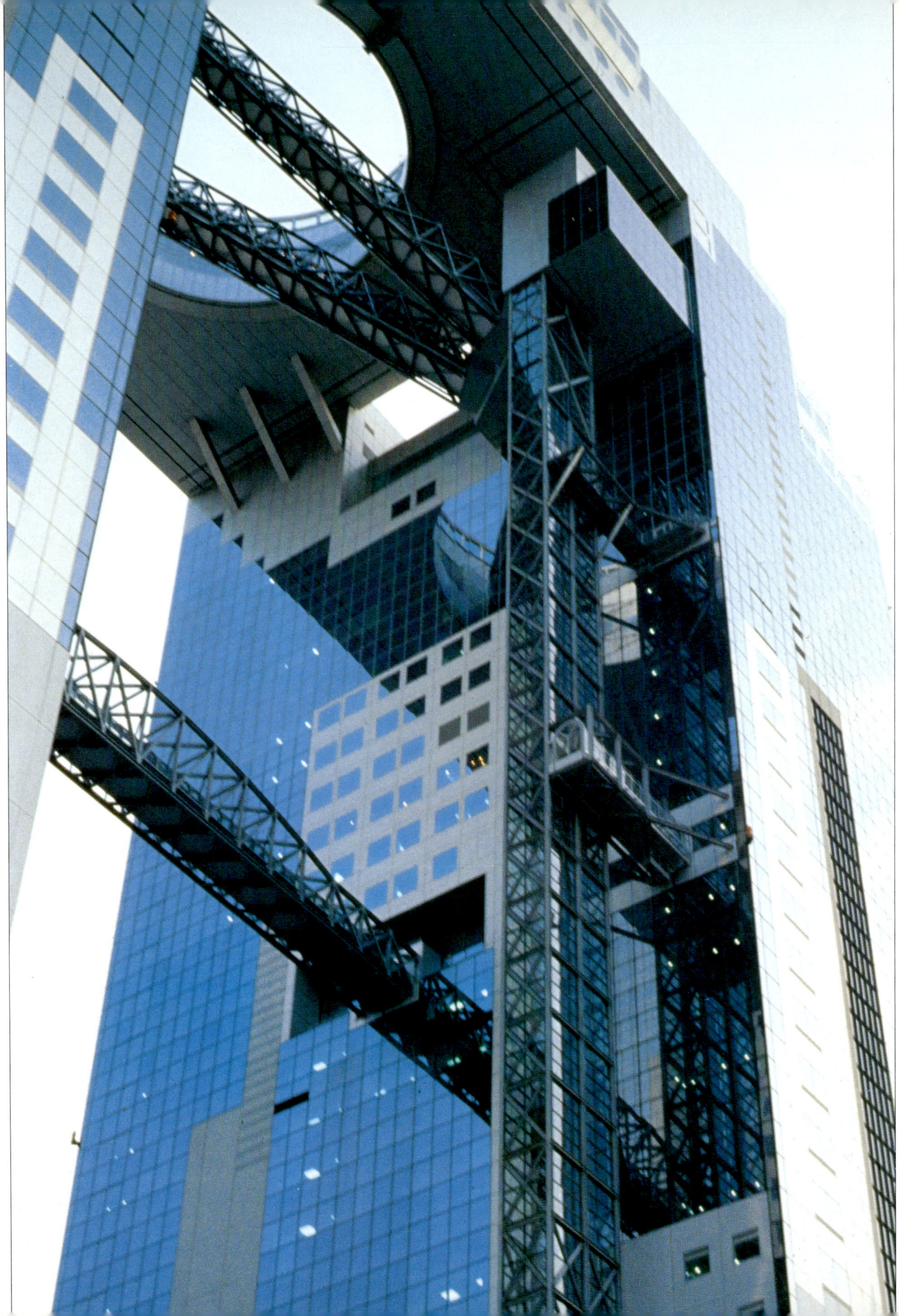

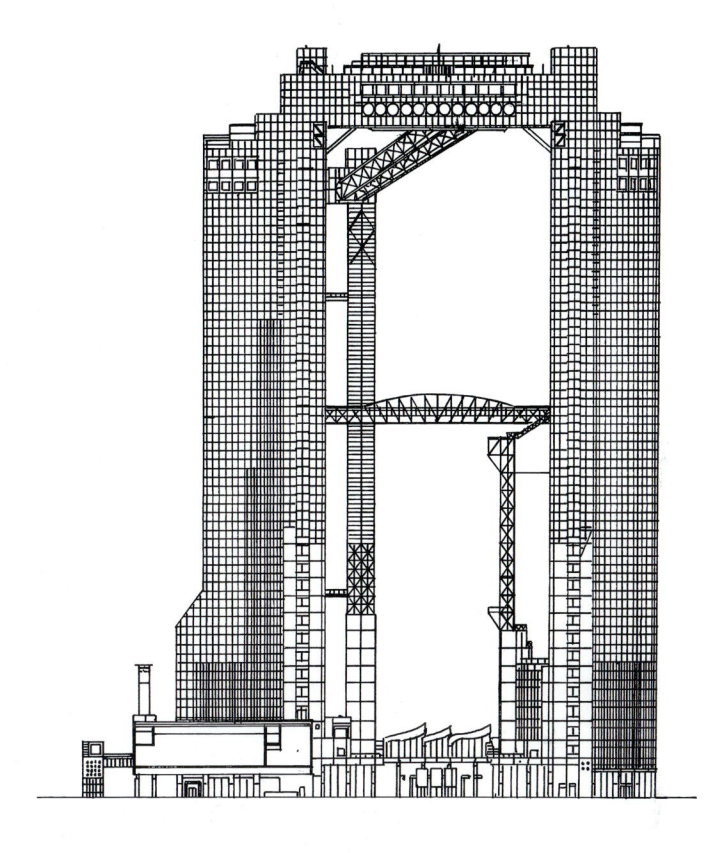

HIROSHI HARA, Umeda Sky Building, Osaka

The completion of this building adds a new and unusual element to the skyline of downtown Osaka. Although the 170 metre structure is one of the tallest in west Japan, this alone does not distinguish it from the ever-growing number of high-rises in the vicinity. It is not one tower but two with approximately 54 metres separation. The buildings are connected in a variety of ways. Firstly, there is the arrangement of underground levels. Secondly, there is a 6-metre-wide bridge on the 22nd level. Finally, there is the three-storey-high structure across the top of the two towers; its size leads the architect to hope that it will be an introduction to the spread of midair cities.

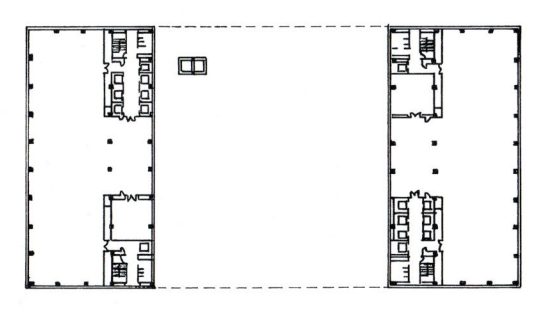

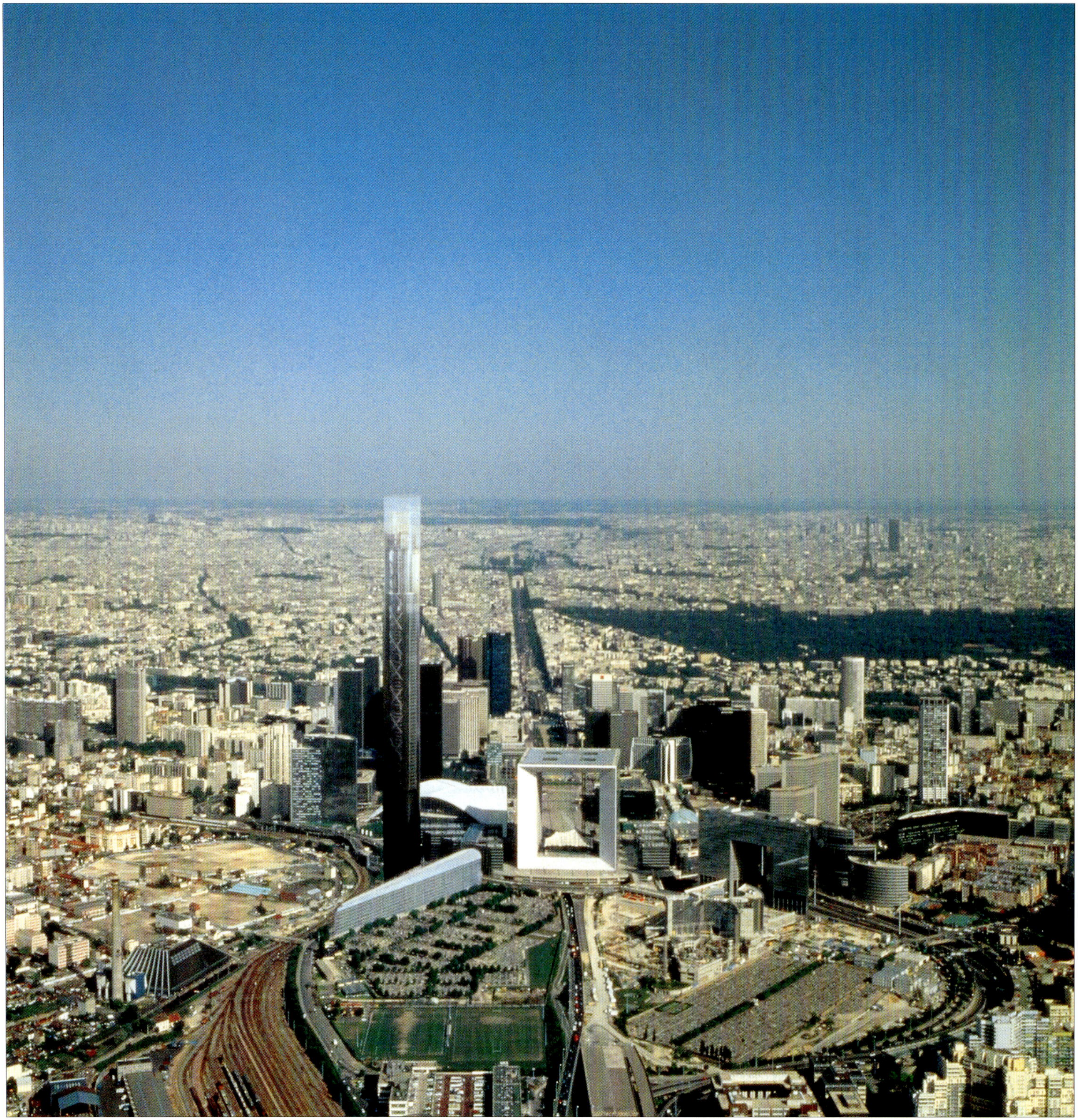

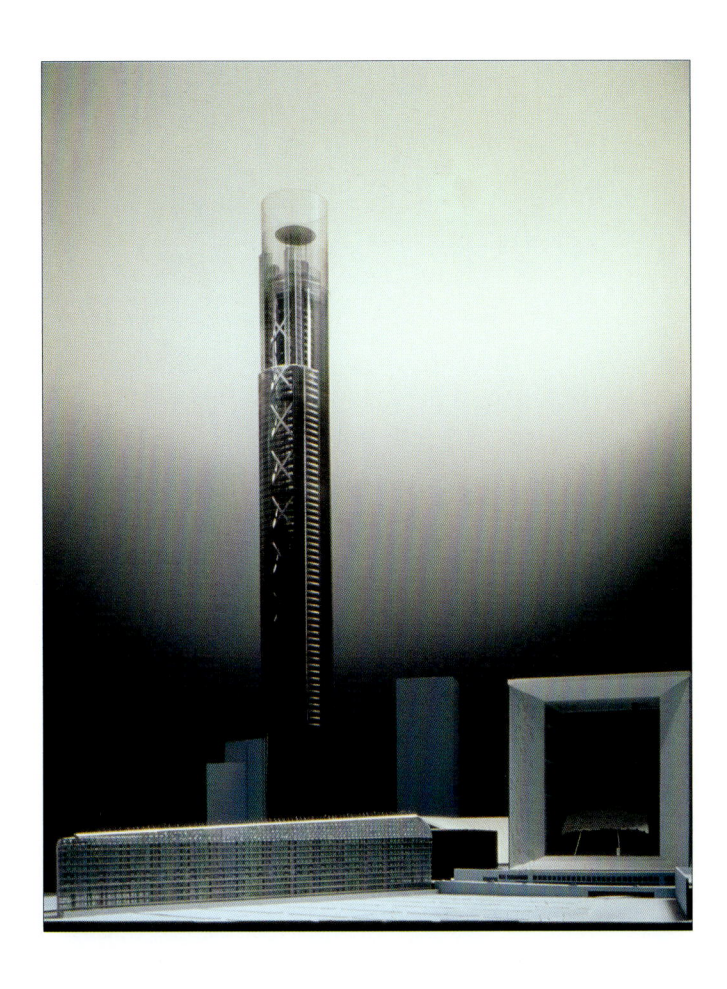

JEAN NOUVEL, Tour Sans Fins, Paris

The tower rises from within the earth, rooted in a crater rather than sitting flat on the ground, a dense granite cylinder that fades away as it reaches to the heavens – a gradual transformation of matter from solid to transparent, evocative of far more than the building's commercial viability and power. This transmutation is a feature not only of the vertical axis but also of the horizontal one, as the building is conceived of as a variety of pleasurable experiences created through subtle variations of natural and artificial light.

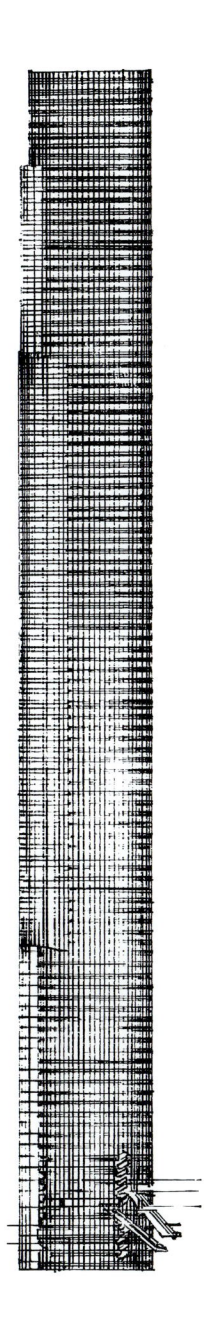

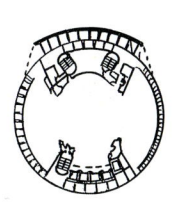

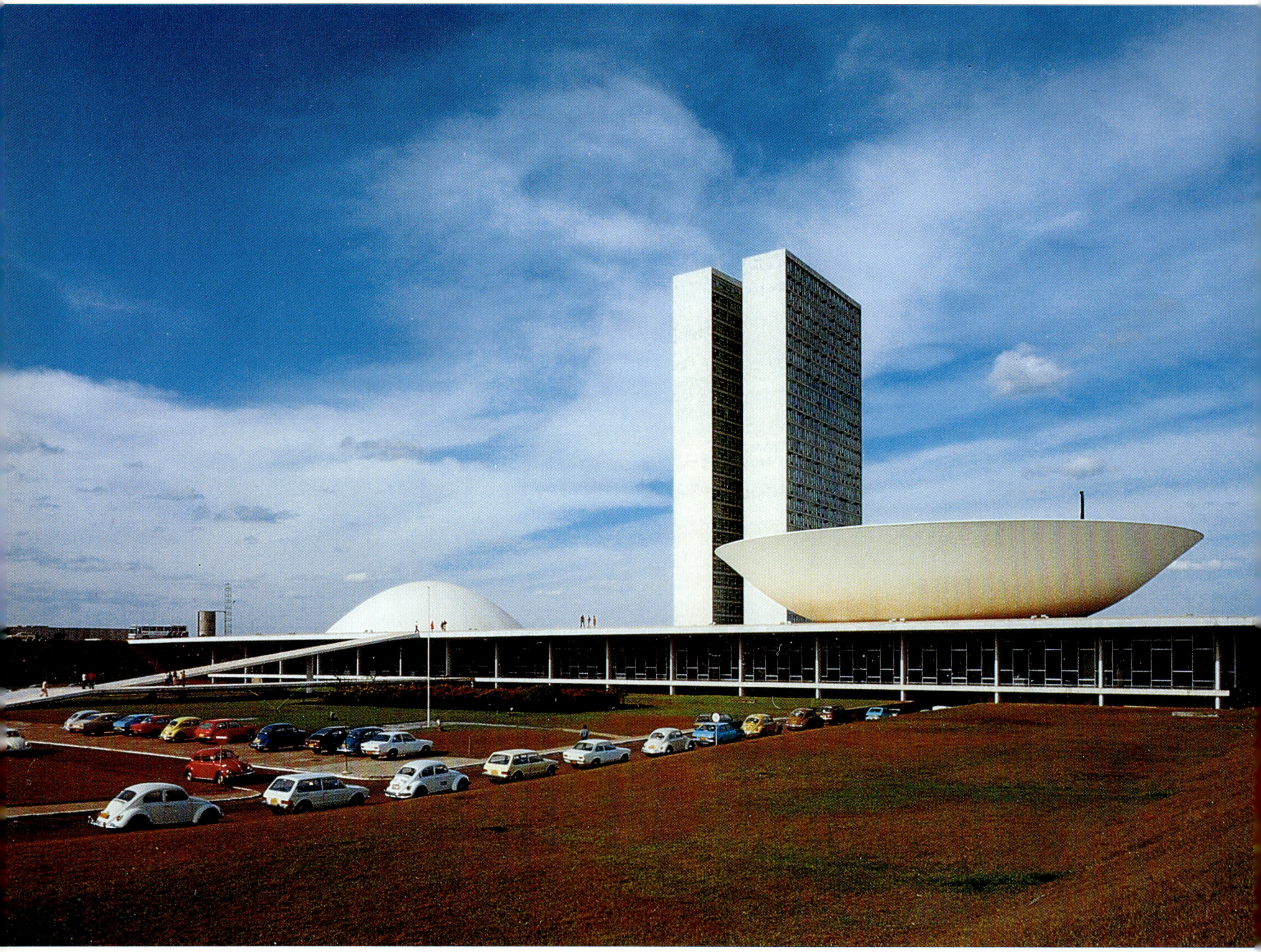

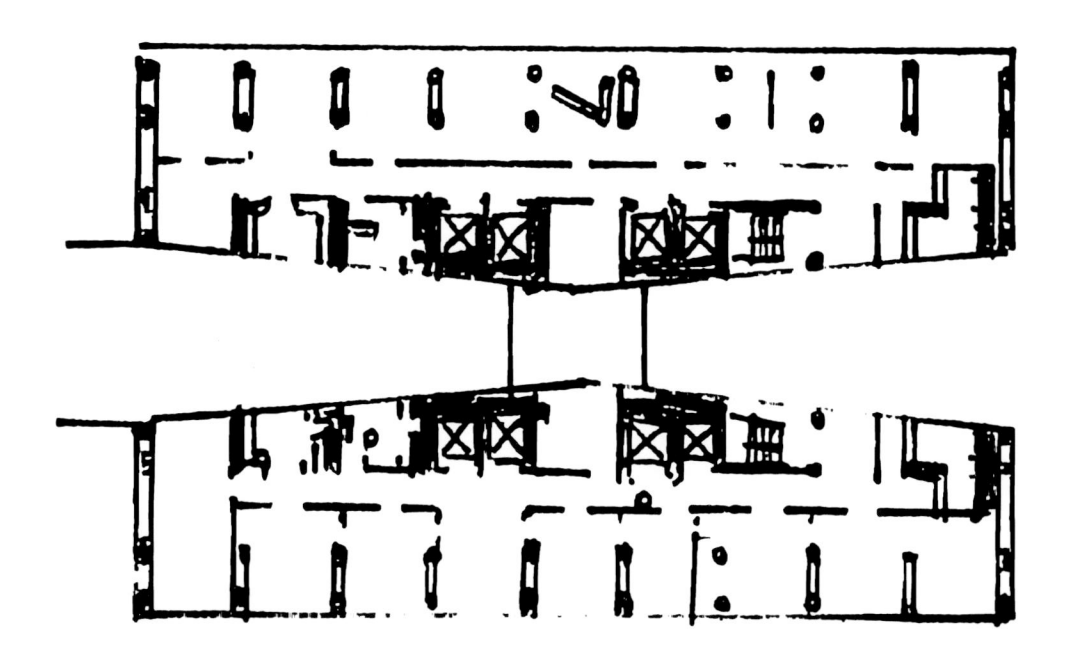

OSCAR NIEMEYER, CONGRESS HALL, Brasilia

After briefly working for Le Corbusier in Niemeyer's home town of Rio de Janeiro in 1936, this world renowned Modernist architect had a string of successes, arguably the height of which was being appointed Chief Architect for the new capital of Brasilia between 1956 and 1964. The forms of the scheme are developed in the beginning in a rather basic manner using everyday objects in order to formulate the composition and then from these simple models the details and the finesse of the scheme are extruded. The local conditions of the project were paramount in the design of the Congress Hall and it has always been a priority for Niemeyer to ensure that the inhabitors of this buildings live in environmental conditions amenable to them. In this way he mixes the caring intentions of an architect designing for the needs of the user, with exact detailing as well as achieving, as an end result, the incredible grand modern architectural gesture.

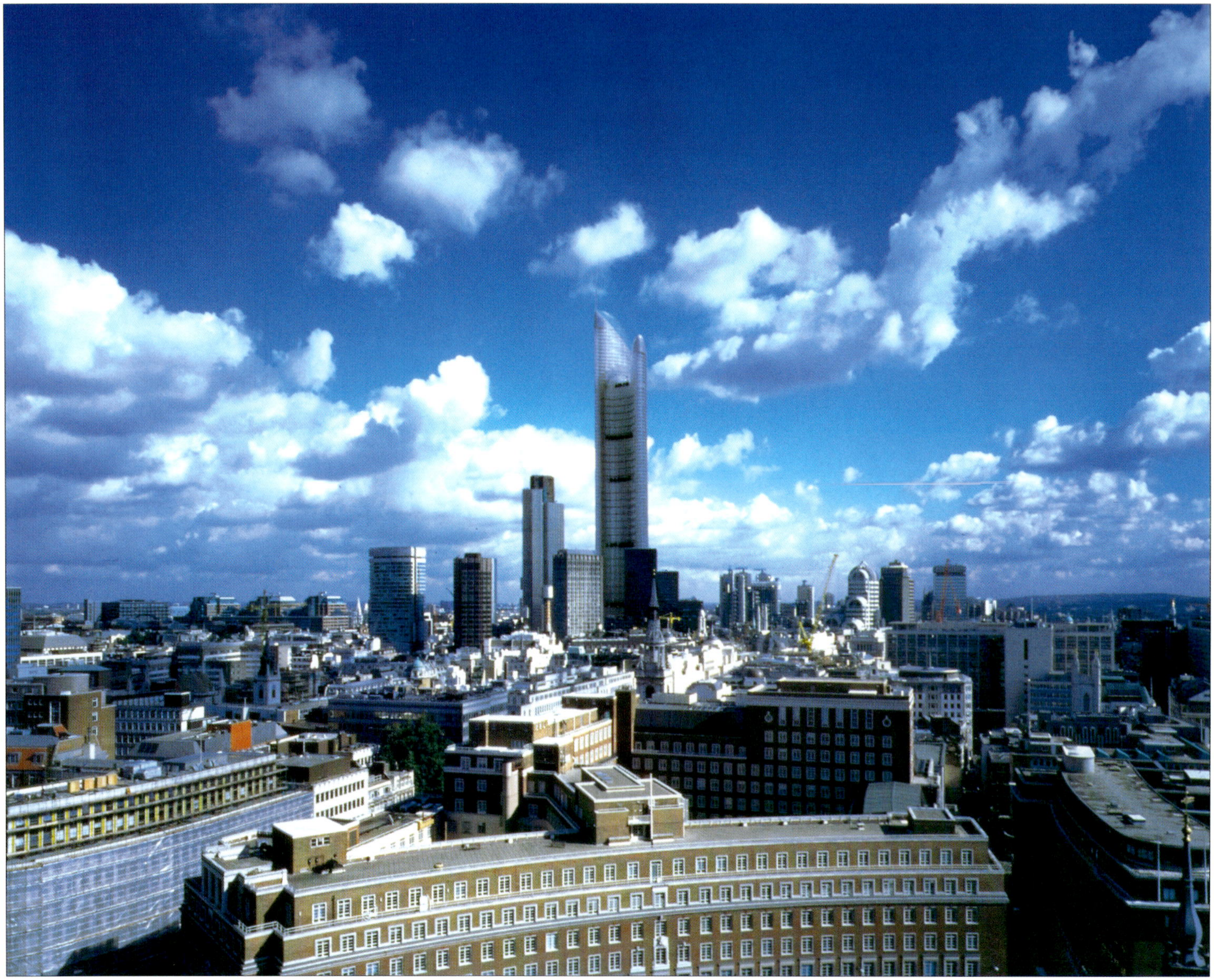

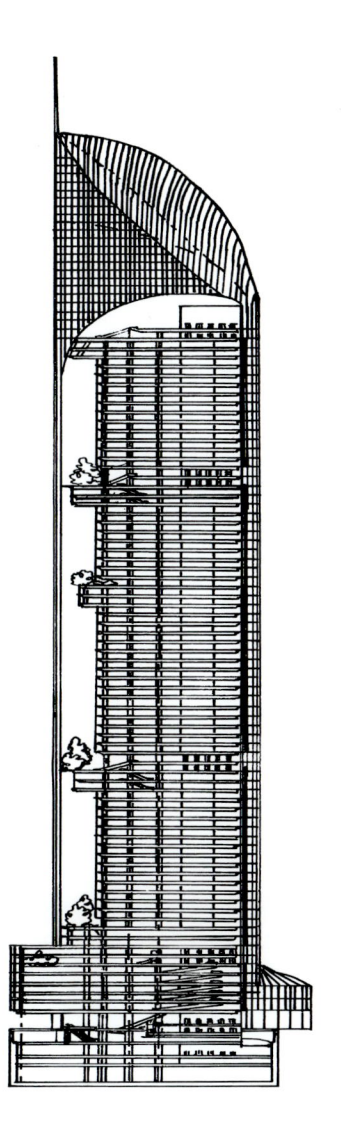

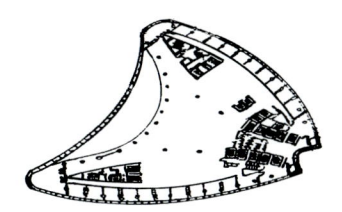

FOSTER AND PARTNERS, Millennium Tower, London

This project was a 92 storey office tower on the site of the Baltic exchange which was badly damaged by an IRA bomb in 1992. At 1,265 feet it would be taller than both the main tower at Canary Wharf and the Commerzbank headquarters in Frankfurt. With a curved free form plan, the building's appearance would constantly change as different qualities of sunlight hit the continuous curves of the glass facade. The top of the building divides into two elegant tail fins of different heights allowing every view of the building to be unique. Views through the glazed double height ground floor lobby, and an open plaza in the front of the tower also create a feeling of spaciousness and light at ground floor level.

FOSTER AND PARTNERS, Commerzbank HQ, Frankfurt

The brief for this tower called for an eco-friendly skyscraper. There are some ingenious innovations in the relatively simple plan which make this a green building. A series of gardens rises up the building in a spiralling pattern providing ventilation and greenery throughout. The plan of the building resembles a chamfered triangle three corners hold the service cores therefore leaving the central triangular atrium core free to be an atrium space that runs up through the entire height of the building.

FOSTER AND PARTNERS, Century Tower, Tokyo

Foster and Partners have moved the 'servant' elements away from the centre of the building in this design leaving the 'served' spaces unobstructed. This immediately opens up much more exciting spatial possibilities, such as the double height floors and central atrium. This device also profoundly affects the external appearance of the building as the service towers become part of the architecture. Consequently the building becomes a collection of towers each with its own architectural expression as opposed to a plain monolith. The Tower has a rich mix of activities. The varied elements are brought together using space and water – still reflecting pools in the public plaza overflow to create vertical cascades down granite walls.

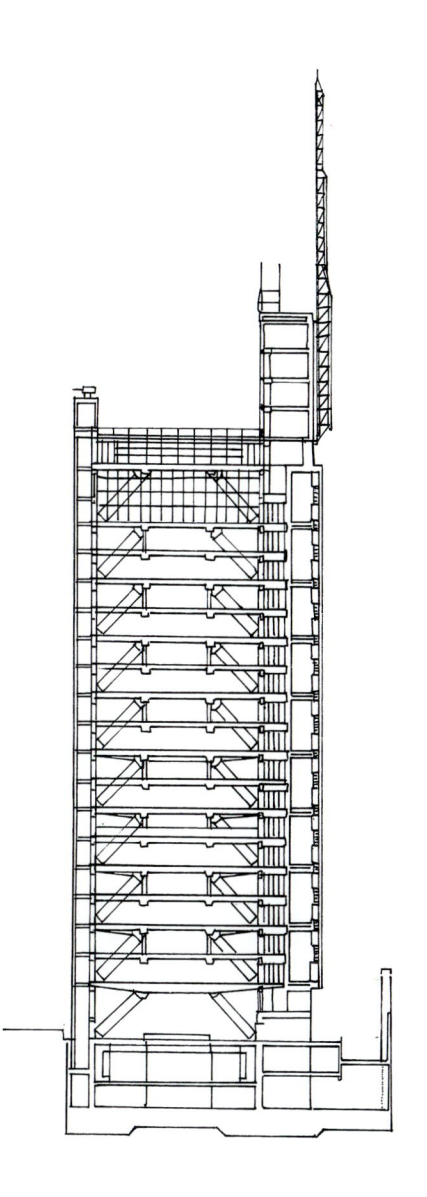

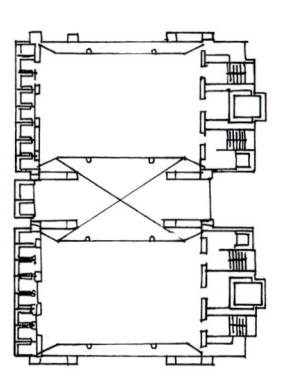

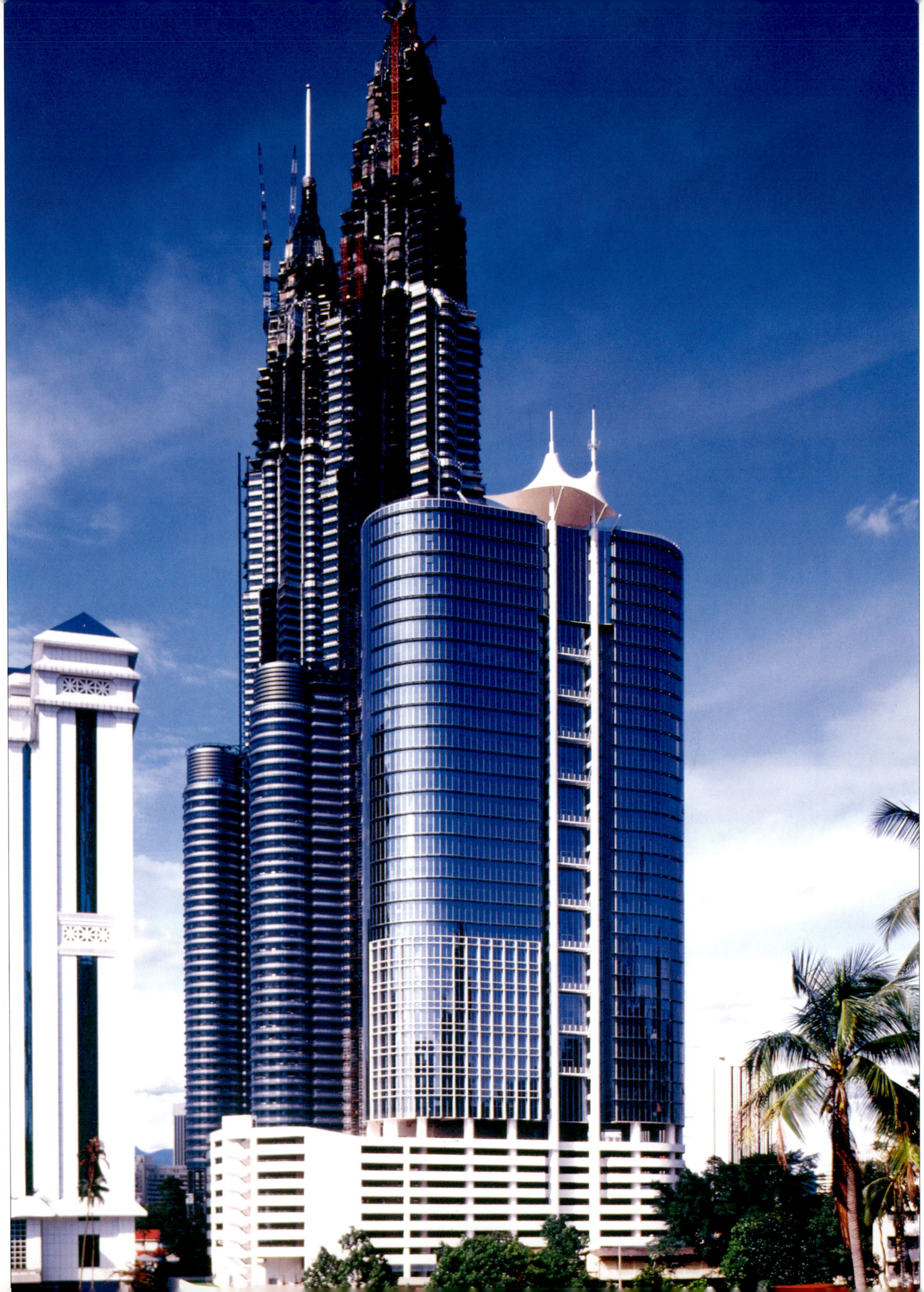

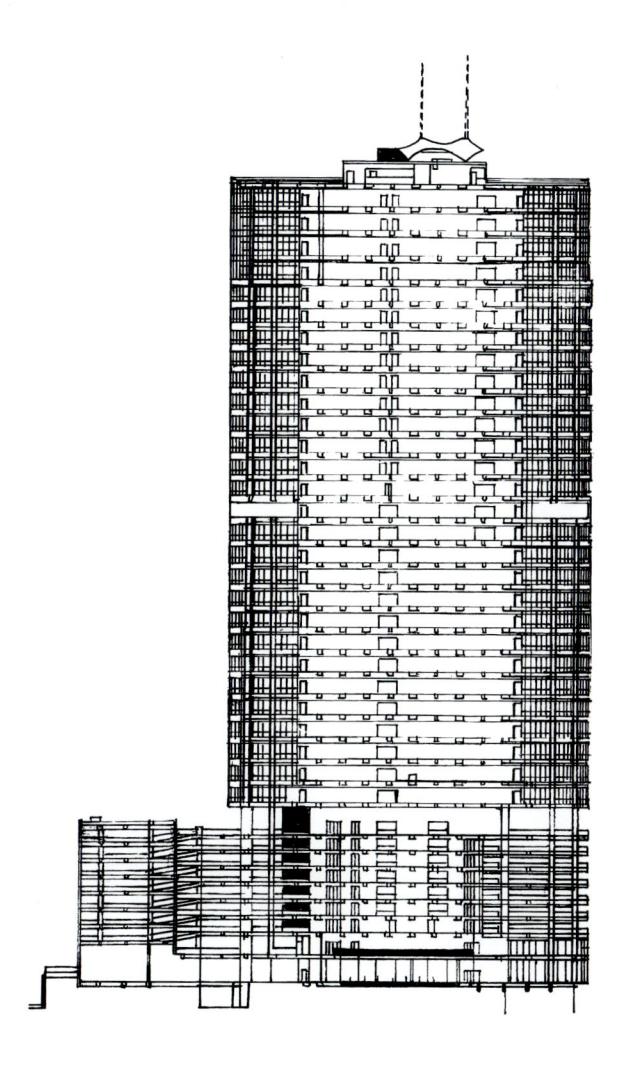

KEN YEANG, Menara TA1, Kuala Lumpur

This rectangular site is orientated diagonally North-South, not an ideal orientation along latitudes near the equator. The site conditions are such that the geometry of the site and of the sun-path do not coincide. The external skin is glazed with a sun-shade system on the West but remains unshielded on the North and South corners where there is minimum solar insulation. The core is located on the East side of the slab, so that the morning sun is kept out of the offices while allowing natural lighting and ventilation into the core areas. The typical internal office floors are column free and on alternate floors open out into a transitional atrium space on the South-West face. The ground entrance lobby is recessed and remains open.

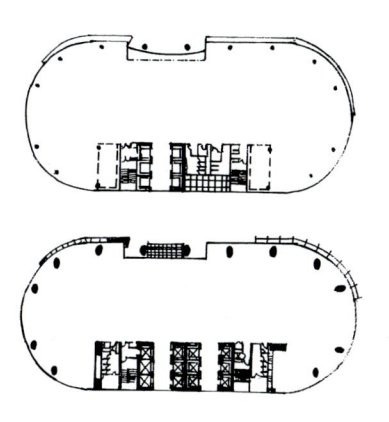

75

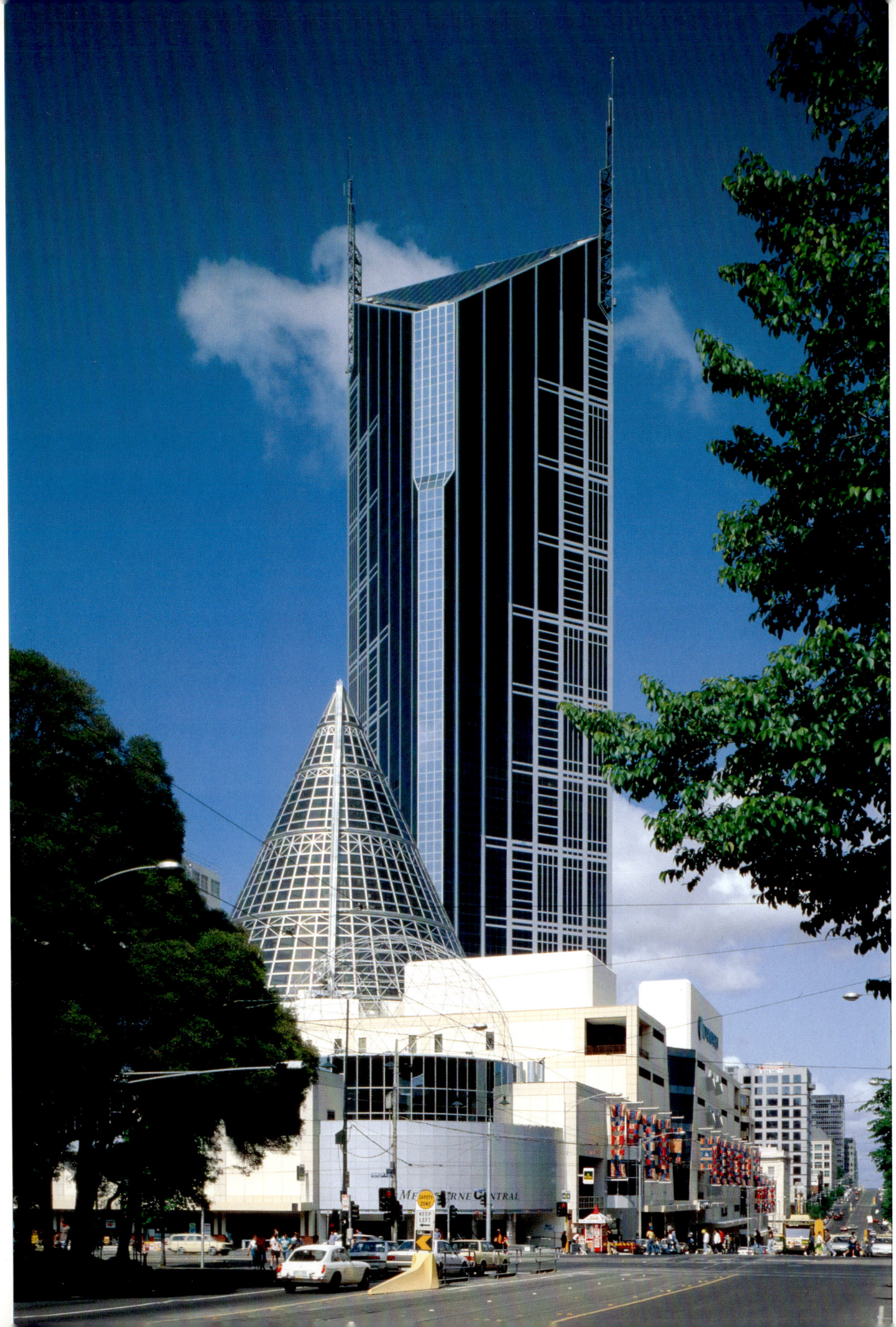

KISHO KUROKAWA, MELBOURNE CENTRAL, Melbourne

The building complex, located in the Melbourne's CBD, is composed of offices, retail space and multiuse entertainment facilities. Inside a large glass cone preserves the existing old Shot Tower. The cone forms an atrium at the centre of the shopping complex, and the relationship between past and present gives the building a feeling of symbiotic coexistence. Within the smooth shape, the facade is a composition of heterogeneous materials, such as stone, aluminum, reflective and tinted glass. High-tech communication equipment is visible at the top of the tower while the lower part of the building is more traditional in design. The facade represents a transition, from the solid city building at the base, which slowly evaporates towards the sky.

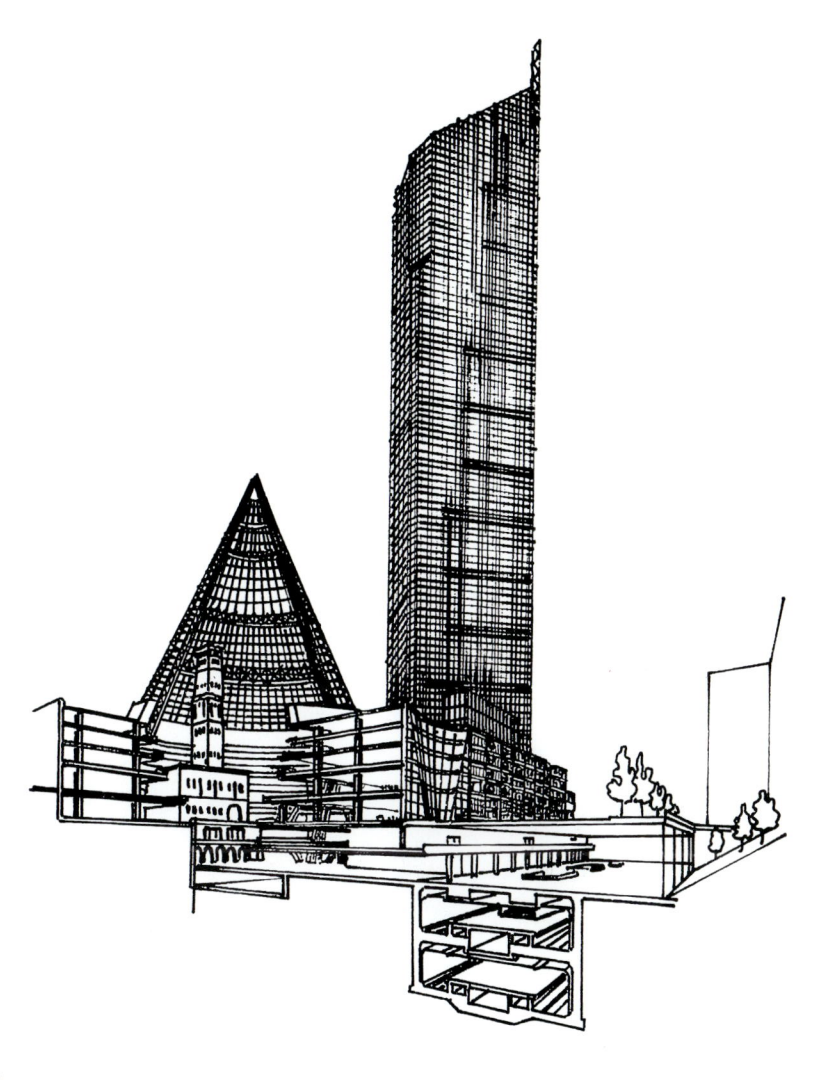

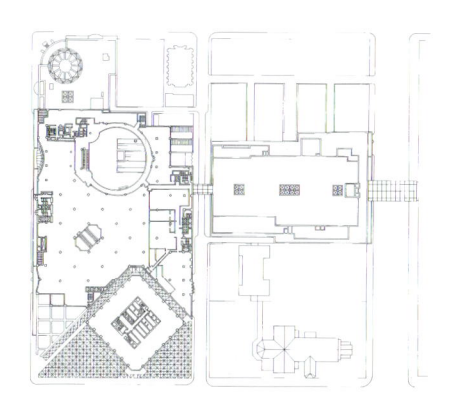

KISHO KUROKAWA, Nakagin Capsule Tower, Tokyo

As residential areas in Tokyo started to shift to the suburbs, this building was a tactical move to restore housing units to the centre of the city, and to provide commuters from outlying areas with studios, bedrooms or a venue for social activities. The individual units were mass-produced and high-tension bolts were used to fasten them to the central core in the desired arrangement. Each room is provided with the facilities of a single hotel room. By replacing or removing the capsules, the appearance of the architecture changes. An evolving architecture contains the potential for participation by the resident in determining its form. By creating spaces of autonomy and individual identity, this building symbolises individual human existence in the urban landscape.

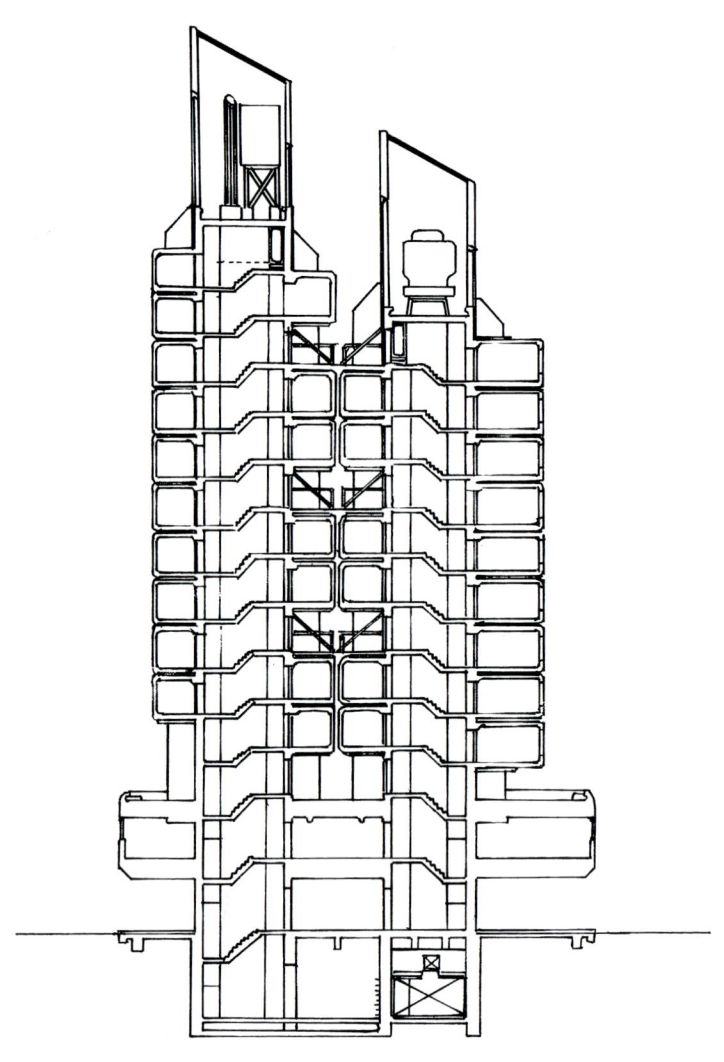

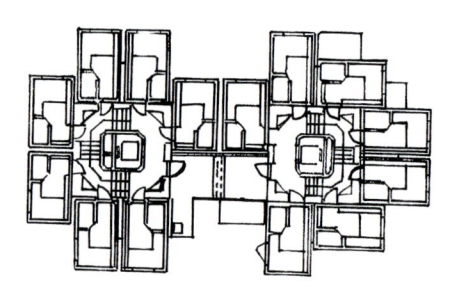

NIHON SEKKEI, Jing Guang Centre, Beijing

This 51 storey high multi-use skyscraper in the central Beijing was built for the Hong Kong Jing Guang Development Co. It houses a five star hotel and office floors as well as luxury appartments on the higher levels and retail stores on the lowest levels at street and below street level. The building covers a site area of over ten thousand square metres and is constructed from a combination of steel, reinforced concrete and steel reinforced concrete. The building was completed in September 1988 when the total floor area of 138,000 square metres began to be used. It has added a much needed resource of property to this area and although it was completed over a decade ago it still looks contemporary.

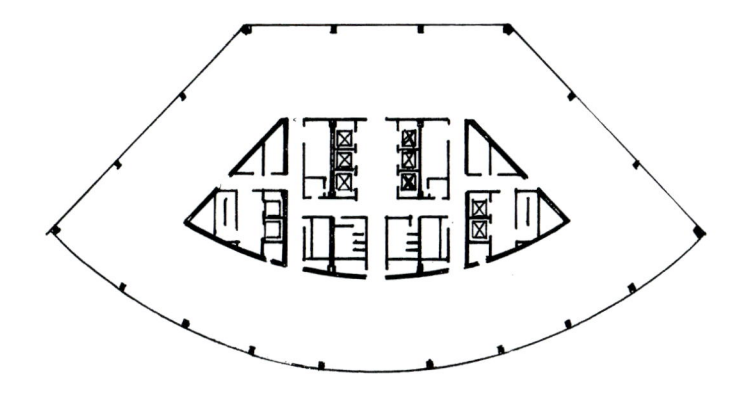

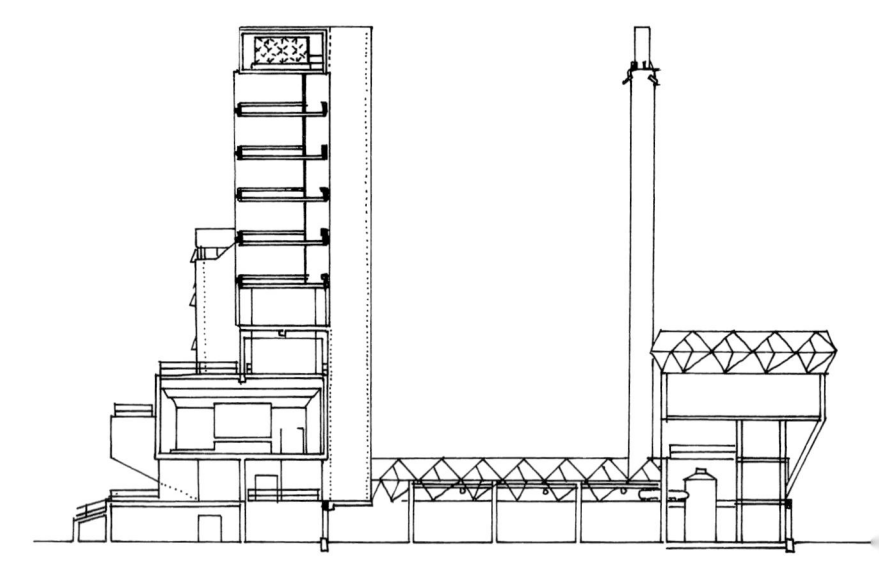

J STIRLING & J GOWAN, LEICESTER UNIVERSITY ENGINEERING BUILDING, Leicester

Completed in 1963, this building was the first that brought James Stirling to public notoriety. The style of the building could perhaps have been anticipated by the firm's design for the Selwyn College completed earlier in the same year and then followed by further developments along a similar theme. Within this building constructed from Stirling's mix of materials including brick and concrete, the design follows the required function of the building. The entire building is serviced through its towers, and these set the precedent for the servicing on the series of buildings completed by Stirling at the beginning of the 1960s.

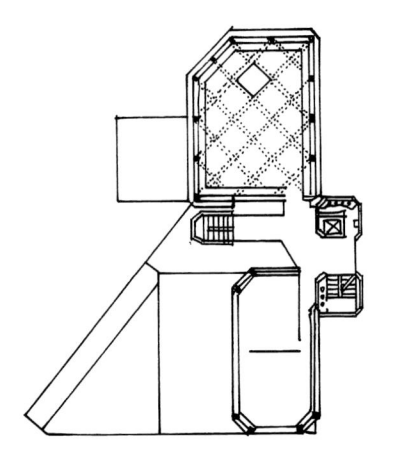

BEN VAN BERKEL, Scuba Towers, London

This study project for part of the old London docks derives from the concept of an architecture of mixed media; an architecture in which foundations are distributed collectively across different zones, and in which the ambition to achieve a spatial, tangible reality of architecture forms a pragmatic point of departure. Mixed media refers to the presence of different elements in the area, of which water is probably the most important. Two of the elements are treated in detail: the water station and scuba tower are both reinterpretations of the various construction techniques used to build the old docks. The water station is a new dock within an existing one, formed by a deep pool dug at the front of the building for the practice of water sports. Hydraulic walls divide the space. The pool is covered with a glass roof at the water level of the old docks which becomes an artificial sheet of water giving a visual indication of the docks, and at the same time interpreting their meaning. The glass roof has an irregular geometric layout deriving from the ground plan. Columns standing on short transverse lines are geared to an undulating longitudinal axis. The city's organisation responds to the arterial meanderings of the river, to a form of movement and activity that is not dependent on programmatic infills.

82

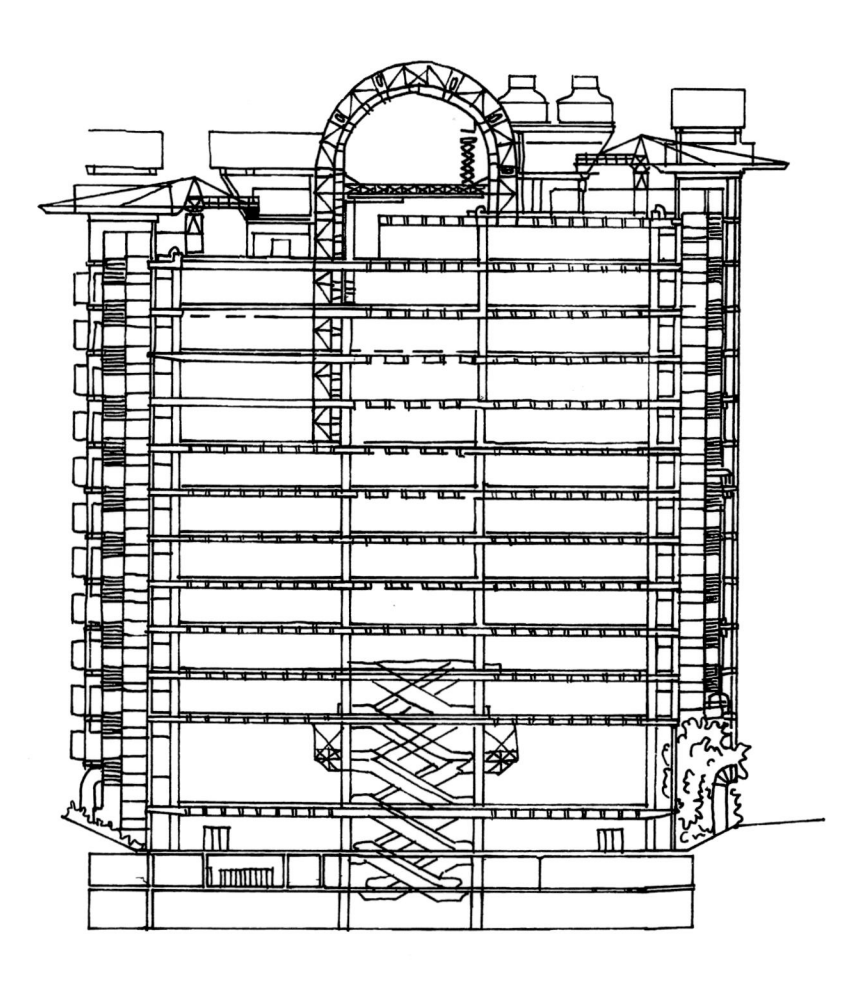

RICHARD ROGERS, Lloyd's Building, London

This scheme is similar in many ways to previous office buildings of a particular type and may well have drawn its inspiration from them. The Larkin Building for example, built by Frank Lloyd Wright, followed a design of open plan and individual offices organised around a covered central courtyard illuminated by natural light. Richard Roger's building at the other end of the century features many aspects from these original ideas, not least of which the open elevator and other features which make the Lloyds Building's design more daring in its structure than its predecessors. Also the numerous advances in ecotechnology used within the servicing of this building potentially enable an improved working environment.

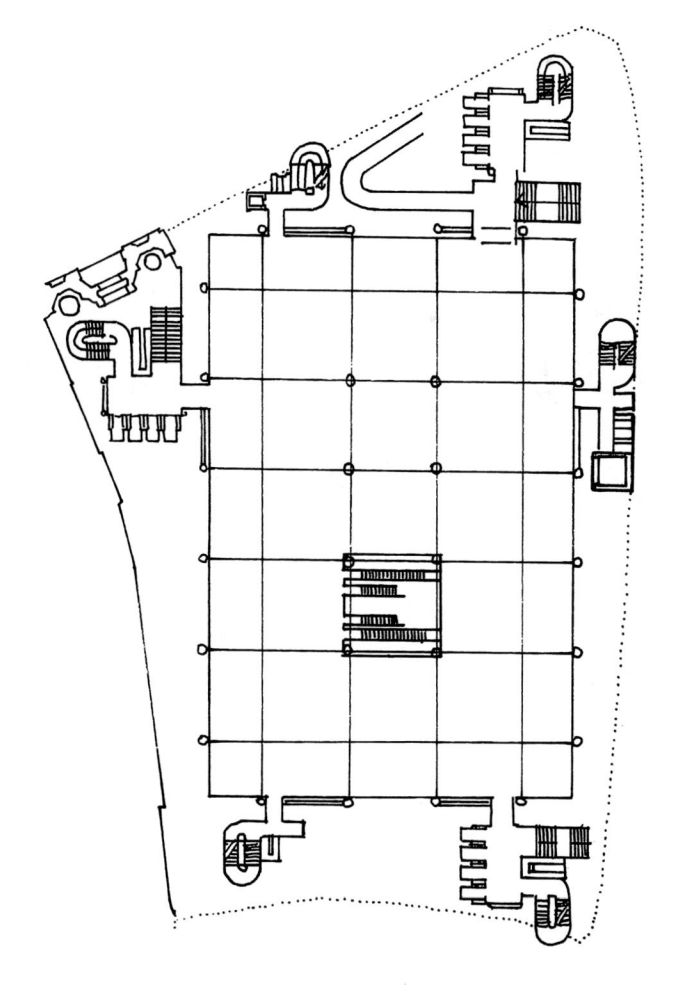

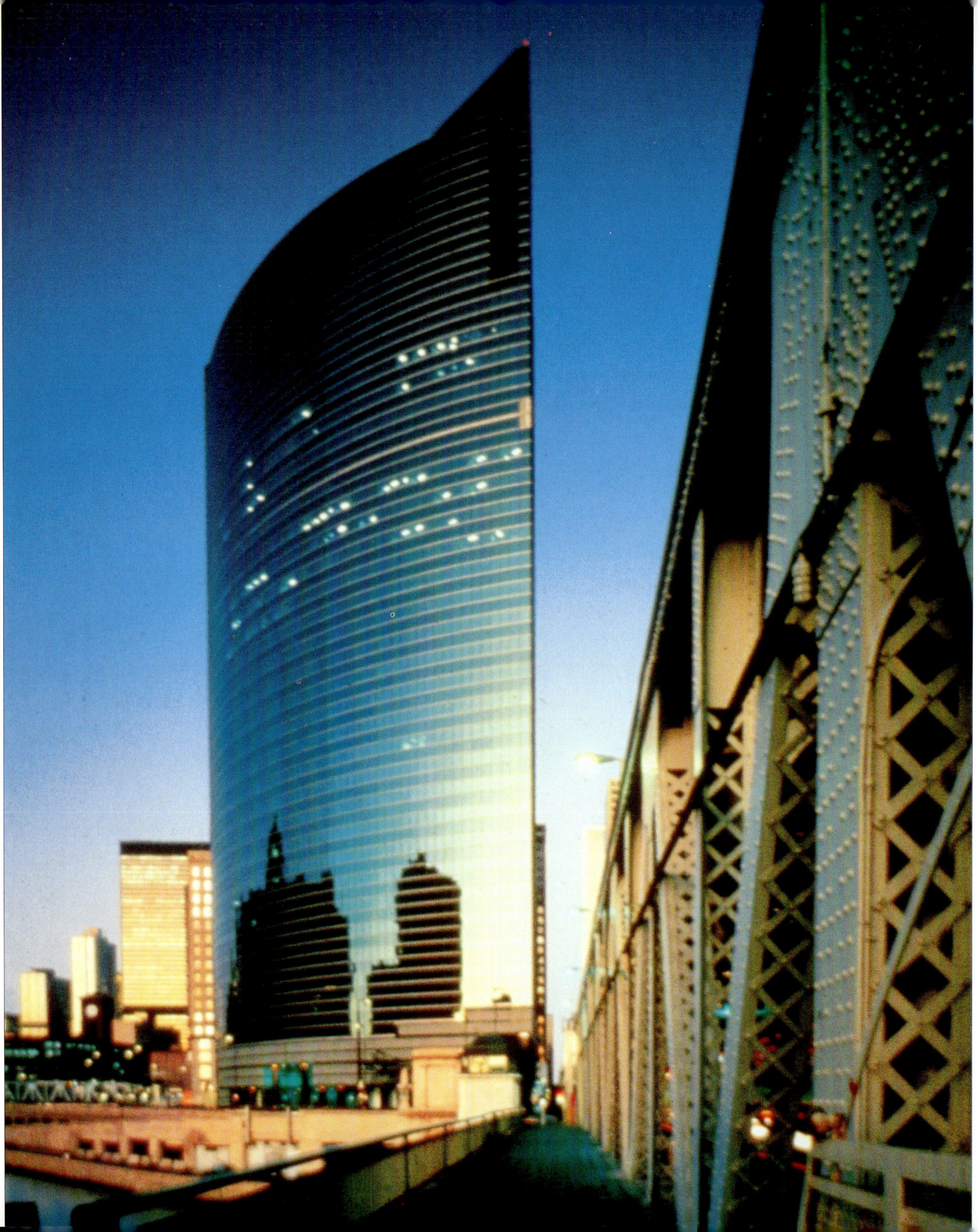

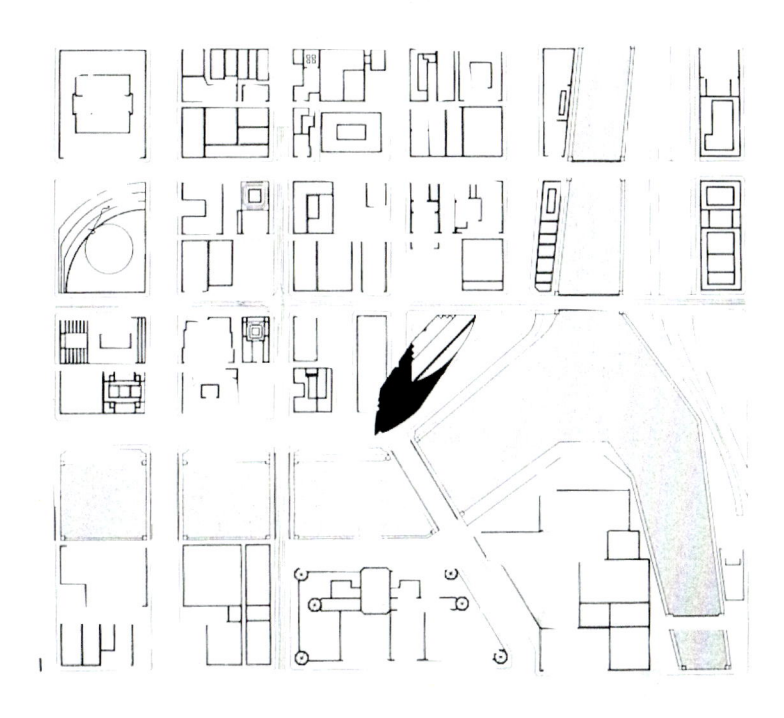

KOHN PEDERSEN FOX, 333 WACKER DRIVE, Chicago, Illinois

Bordered by South Wacker Drive and Franklin and Lake Streets, this award-winning 36-storey, 100,000 square metre office tower is situated on a tight triangular plot along the Chicago. The building's riverside facade presents a taut curve dramatised by a sleek, reflective green glass curtain wall reinforced with horizontal bull noses at 6 foot intervals. Supported by a horizontally banded grey granite and green marble base which encloses a large mechanical floor, the first office floor sits above the EL (elevated train) where it passes the site along Lake Street. The building's downtown facade echoes the geometric city grid and opens into a two-storey interior lobby coolly accented with terrazzo floors and stainless steel trim.

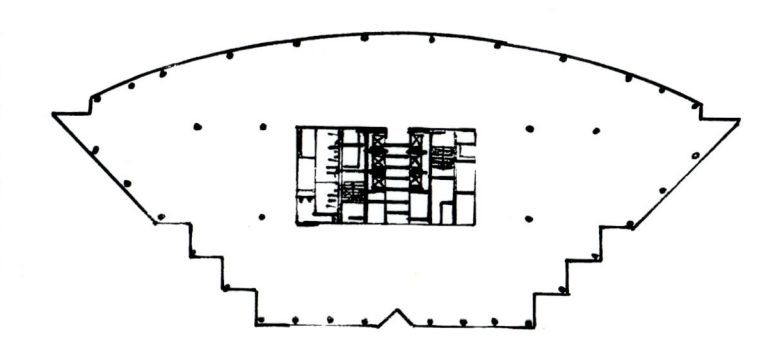

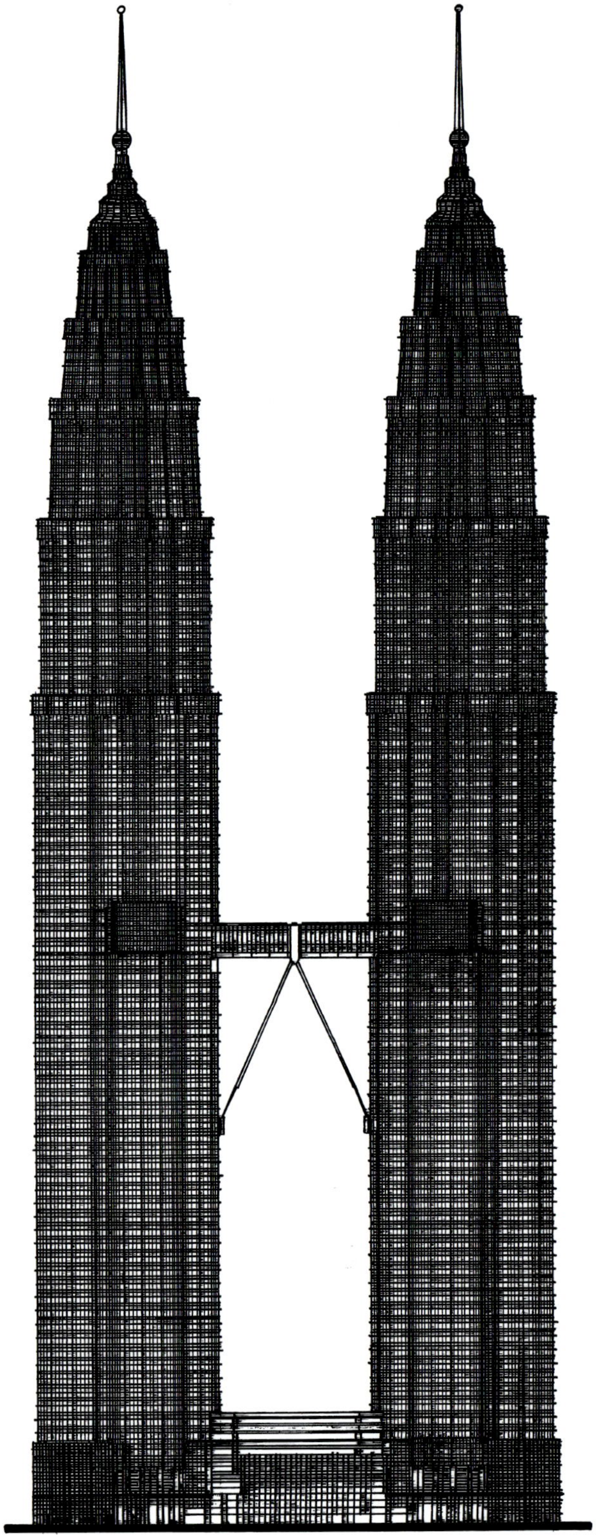

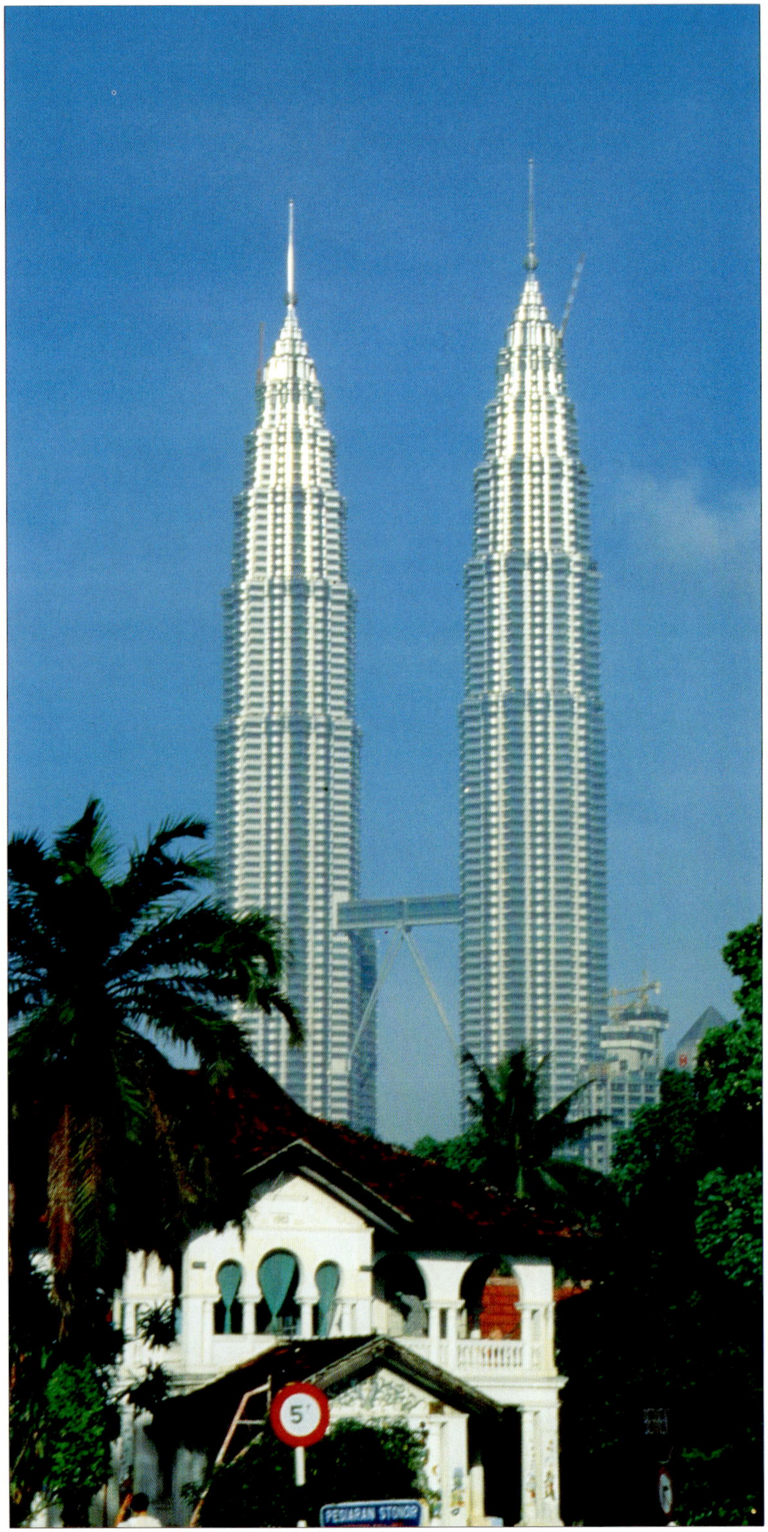

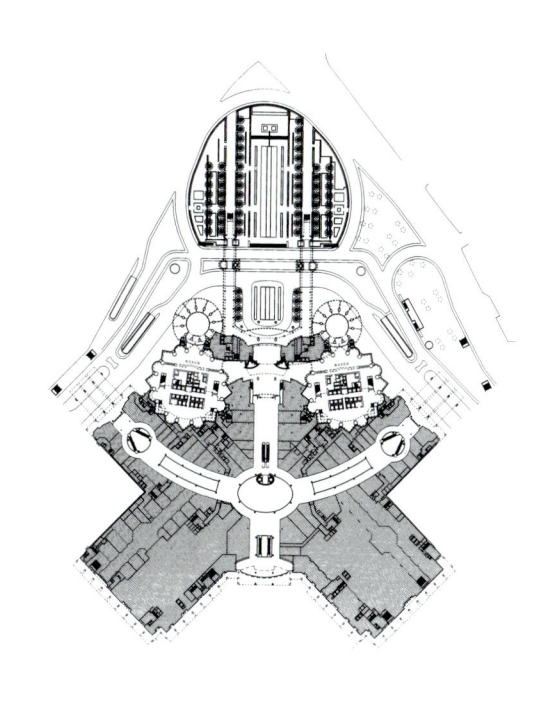

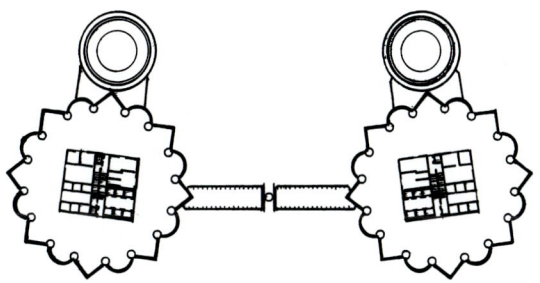

CESAR PELLI, Petronas Towers, Kuala Lumpur

To meet the demand of urban growth in Malaysia's federal capital, Kuala Lumpur, the government has decided to allow the Selangor Turf club and its surrounding land, strategically located in the heart of the commercial district, to be developed into a new 'city within a city' that will demonstrate the ideal working, living and recreational environment in a parkland setting. At the centre of this development stand the twin towers, designed to be both functional and beautiful as well as defining an urban gateway of monumental scale with their distinctive 'skybridge'. At a height of 451.9 metres the towers are recognised as one of the world's tallest structures.

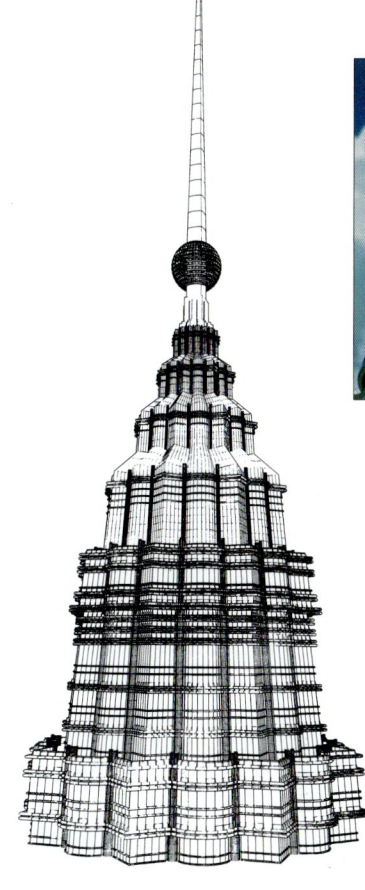

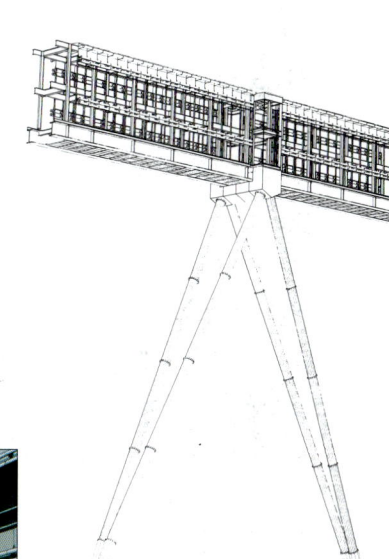

CHECKLIST OF M&E SERVICES (FOR INTEGRATION WITH SERVICE CORES)

Item	Integration with Service Core		
	Crucial	Important	Unimportant
HVAC			
Fuel Sources			
• Gas			√
• Oil			√
• Solar			√
• Waste			√
• Electric grid			√
• Electric, district or local generation			√
• Other			
Service Generators, Boilers/Furnace			
• Hot water		√	
• Low-pressure steam		√	
• High-pressure steam		√	
• Fire tube		√	
• Water tube		√	
• Cast iron		√	
• Steel		√	
• Gravity warm air			
• Forced warm air			
• Central or distributed; # per sq ft, or per population density			
Compressors/Furnace			
• Refrigeration		√	
• Electric drive		√	
• Absorption units		√	
• Reverse thermosiphon		√	
• Central or distributed; # per sq ft, or per population density		√	
Central Air Conditioning			
• Cooling tower motor		√	
• Air-cooled condenser motor		√	
• Condenser pumps		√	
• Room air conditioners		√	
• Through-the-wall units		√	
• Economiser cycle		√	
Ventilation/Fresh Air			
• Minimum settings		√	
• Economiser		√	
• CO$_2$ / pollution sensors/controllers		√	
• Intake/exhaust configuration		√	
• Distributed systems	√		
Pumps (# units, connected HP)			
• Chilled-water pumps		√	
• Condenser water pumps		√	
• Boiler feed pumps		√	
• Hot-water pumps for space heating		√	

Item	Integration with Service Core		
	Crucial	Important	Unimportant
• Recirculating pumps for domestic hot water			√
SERVICE CONDUITS			
All-Air-HVAC Systems (# of a.h.u., total HP cfm/a.h.u.)			
• Single zone			√
• Terminal reheat (hot water electric, steam)			√
• Variable air volume	√		
• Induction		√	
• Dual duct	√		
• Multizone units		√	
• Unitary heat pumps			√
• Sq ft vertical plenum/total sq ft			√
• Depth horiz plenum/slab-to-slab height			√
Water-Air Systems ("# units, connected HP")			
• Two-pipe fan coil			√
• Three-pipe fan coil			√
• Four-pipe fan coil	√		
• Unitary heat pumps			√
Service Terminals ("# units, connected HP," cfm/fan unit)			
Fans (supply and exhaust)			
• Backward-curved multivane fans			√
• Forward-curved multivane fans			√
• Axial fans			√
• Propeller fans			√
Perimeter Units			
• Fin-tube radiators			√
• Cast-iron radiators			√
• Radiant heating coils			√
• Hot-water piping			√
• Supply and return ducts			√
• Outside air dampers			√
Local Distribution			
• Ceiling, floor, wall or furniture			√
• Density diffusers, per sq ft per population			√
• Air flow			√
• Diffuser shape, configuration			√
• Material, ornament			√
• Access			√
• Interface/Expansion			√
• Relocation capability			√
Systems Integration			
• Structure			√
• Interior ceiling			√
• Interior floor			√
• Interior partitioning			√

Item	Crucial	Important	Unimportant
• Lighting			√
• Enclosure (MRT management, perimeter/core balancing)			√

ENERGY MANAGEMENT CONTROL SYSTEM (EMCS)

Central Management

Item	Crucial	Important	Unimportant
• Manual		√	
• Automatic		√	
• Location		√	
• Grouped controls and displays		√	
• Major functions		√	
• Indicators and controls		√	
• Schema 'as wired'/'as installed'		√	
• Emergency telephone with separate line (not via PBX or switchboard)		√	
• Tests conducted		√	
• Scheduled maintenance		√	
• Performance/evaluation annually		√	
• Performance/evaluation semiannually		√	

Resource Management Systems

Item	Crucial	Important	Unimportant
• Thermal storage (water, ice, other)		√	
• Peak power sharing		√	
• System shutdown/setback		√	
• Individual sensors		√	
• Programmable sensors		√	
• Density/type of sensors (temp, CO_2 particulates, humidity)		√	

Local Management Systems/Controls

Item	Crucial	Important	Unimportant
• Location		√	
• Automatic/Manual		√	
• Density		√	
• Individual HVAC systems		√	
• Radiant panel		√	
• Terminal temperature control		√	
• Terminal air control		√	

Utility-Energy Data Electricity

Item	Crucial	Important	Unimportant
• KWH/month/per service (i.e. lighting, air conditioning, hot water, office equipment)		√	
• Peak demand/month		√	
• Connected load		√	
• KWH/sq metre/year		√	

Oil

Item	Crucial	Important	Unimportant
• BTU/gallon		√	
• Gallons/month		√	
• BTU/sq metre/year		√	

Gas

Item	Crucial	Important	Unimportant
• BTU/Cu ft		√	
• Cu ft/month		√	
• BTU/Cu ft/month		√	

Obtain

Item	Crucial	Important	Unimportant
• Electric rate schedule		√	
• Fuel; oil and gas rate schedule		√	

To Determine

Item	Crucial	Important	Unimportant
• HVAC – if VAC, how are fans controlled?		√	
• Vaned inlet, multispeed motor, frequency controller, other			√
• Heat recovery – sources, applications			√
• Vibration control in building for instrumentation			√
• Wet and dry waste management			√
• Water-saving devices			√
• Energy storage – ice-chilled water – latent heat			√

Climate Forecasting

Item	Crucial	Important	Unimportant
• Max high/month			√
• Average high/month			√
• Average temperature/month			√
• Average low/month			√
• Average max/month			√
• Wind velocity – direction/intensity			√
• Solar radiation global/month (direct, diffuse, cloud cover, sunshine)			√
• Design wet bulb and dry bulb			√

Occupancy Density

Item	Crucial	Important	Unimportant
• Workers	√		
• Visitors	√		

EMCS

Item	Crucial	Important	Unimportant
• Is there energy optimisation?			√
• Is indoor–outdoor air quality monitored and controlled?			√

Plumbing Configuration

Item	Crucial	Important	Unimportant
• Size volume			√
• Form, configuration	√		
• Planning module	√		
• Expansion capability		√	
• Material, ornament			√

Method of HDW Generation and Storage

Item	Crucial	Important	Unimportant
• Oil, gas, electricity, coal for DHW			√
• Tankless heater on space-heating boiler			√
• Tank heater on space-heating boiler			√
• Tank insulation thickness			√

Service Conduits/Piping

Item	Crucial	Important	Unimportant
• Thickness, volume	√		
• Configuration, distance			√
• Interface, expansion			√
• Material, ornament			√
• Access		√	

Fixtures in Service Terminals (kitchen, toilets, etc)

Item	Crucial	Important	Unimportant
• Planning module	√		
• Number, size, capacity	√		
• Form, ergonomics		√	
• Material, ornament			√
• Integration with interior walls, ceiling			√
• Interface/expansion		√	
• Relocation capability			√

Item	Integration with Service Core		
	Crucial	Important	Unimportant
• Access and maintenance		√	
Conservation			
• Grey water system			√
• Water conservation strategies			√
• Secondary use of stored water (off-peak, fire)			√
Fire-safety Service Generators			
• Wet			√
• Dry			√
• Delayed action			√
• Halon 1301			√
• Manual pull stations			√
• Pre-action sprinklers			√
Planning Module			
• Size, capacity		√	
• Zoning		√	
• Raised floors		√	
• Closets		√	
• Ceiling – hidden			√
• Exposed			√
• Access, maintenance		√	
Sensor/Controllers			
• Optical/photoelectric			√
• Smoke			√
• Heat/temperature			√
• Ionization			√
• Flame			√
• HVAC modifications			√
• Local shut off/delay			√
• Halon discharge controllers			√
• Pre-alarms			√
• Time delays			√
• Leakage 'tattletales', water monitoring			√
• Cross-zone alarms			√
Integration			
• Structure			√
• Interior ceiling			√
• Mechanical/HVAC			√
• Lighting			√
• Interior partitioning			√
Automatic Systems			
• Central computer			√
• HVAC			√
• Automatic start-up fans			√
• Automatic control fire damper			√
• Air-conditioner shutdown			√
• Blocking air ventilation/positive and negative pressurisation			√
• Control access computer			√
• Life-safety control system			√

Item	Integration with Service Core		
	Crucial	Important	Unimportant
• Manual and automatic alarm systems			√
• Separate fire-related circuits			√
• Data log identification			√
• Computer room doors opening			√
• Easy restart of all systems			√
Disaster Resistance			
• Equipment shutdown			√
• Power shut off 'panic' button			√
• Halogen gas			√
• Alarms annunciate at control centre			√
• Fire dampers			√
• Emergency lighting			√
• Life-safety system			√

HVAC NOTABLE OVERALL PERFORMANCE

	Crucial	Important	Unimportant
Spatial Quality			√
Acoustic Quality			√
Air Quality			√
Thermal Quality			√
Visual Quality			√
Building Integrity			√

NOTABLE SYSTEMS INTEGRATION
HVAC and Envelope, HVAC and Lighting, HVAC and Power/Telec, HVAC and Vertical Transportation, HVAC and Structure, HVAC and Interior

Item	Crucial	Important	Unimportant
• Air-flow window			√
• Light shelf ducts			√
• Shared plenum/grid management			√
• Shared core planning; high accessibility	√		
• Use of structure for HVAC			√
• Personal environmental controls, workstation harness			√

PLEC (Power, Lighting, Electronics and Communication) Systems, Transport and Security System
Service Generator
Central Power Type

	Crucial	Important	Unimportant
• Utility power – single – or multiple-grid supply			√
• Self-contained generation			√
• Cogeneration			√
• Emergency standby power			√
• Uninterrupted power supply (UPS)			√
• Clean power vs general power separation			√

Size (KVA)

	Crucial	Important	Unimportant
• Standard building load			√
• Oversized (automation growth)			√

Capacity Planning Module

	Crucial	Important	Unimportant
• 110V / 120V			√
• 208V / 277V / 480V			√

Item	Integration with Service Core		
	Crucial	Important	Unimportant
• 50 Hz / 60 Hz			√
• 400 Hz			√
• Emergency stand-by			√
• Diverse utility feeds			√
• Uninterruptible power system			√
• Motor generator			√
• Automatic power transfer			√
• Power conditioning to terminals			√

Location/Distribution

Item	Crucial	Important	Unimportant
• Basement utility room			√
• Roof			√
• Bullet middle floors			√

CENTRAL DATA/VIDEO
Internal Communications Systems

Item	Crucial	Important	Unimportant
• Voice-mail system			√
• Electronic mail			√
• Videotex			√
• Internal FAX			√
• Communicating copiers			√
• Internal videoconferencing			√

Computer System

Item	Crucial	Important	Unimportant
• Mainframe/multiprocessors			√
• Supercomputer (Cray, etc)			√
• Minis/networks and gateways			√
• Micros/networks and gateways			√

External Communication Systems (ECS)

Item	Crucial	Important	Unimportant
• Microwave			√
• Satellite			√
• Access to packet-switched network			√
• Elect tandem network (ETNs)			√
• Facsimile, teletype, telex			√
• Videoconferencing			√
• Teleconferencing			√
• Teleport			√
• RF systems			√
• CATV			√
• Transparent internal gateways to ECS			√

Shared Tenant Systems

Item	Crucial	Important	Unimportant
• Full building supported		√	
• Bullet partial building shared		√	
• Realcom (IBM)		√	
• Contel		√	
• TelCom Plus		√	
• AmeriStm (Lincoln Properties)		√	
• Martnet (Trammell Crow)		√	

CENTRAL TELEPHONE
Type

Item	Crucial	Important	Unimportant
• Central-office based (CENTREX)			√
• Private branch exchange (PBX)			√
• Key systems			√
• Digital service from central office (ISDN)			√

Telephone Service Entrance Size and Facilities

Item	Crucial	Important	Unimportant
• 1–1 number of telephones			√
• Less than 1-1 (internal switch)			√
• % of telephone anticipated			√
• Direct Inward Dailing (DID)			√
• Fibre-optic building service			√
• Central office trunking over T–1			√
• Spans to PBX			√
• Multiplexers in building			√

Main Telephone Room Location Size

Item	Crucial	Important	Unimportant
• Close to riser distribution		√	
• Convenient to service entrance			√
• To house existing equipment			√
• Planned expansion			√
• Environmentally controlled			√

POWER, DATA AND VOICE
Vertical Distribution

Item	Crucial	Important	Unimportant
• Sufficient density			
• Empty riser space for expansion/new cabling			
• Distributed risers, number and location		√	
• In conduits and riser ducts separate from power conductors	√		
• Separated and/or shielded from sources			

Network Topology

Item	Crucial	Important	Unimportant
• Star			√
• Bus			√
• Ring			√
• Hybrid			√

Premise Wiring Scheme

Item	Crucial	Important	Unimportant
• AT&T PDS			√
• IBM			√
• DEC			√

Networking

Item	Crucial	Important	Unimportant
• Local area networks			√
• LAN compatible computer equipment			√
• LAN variety of vendor equipment			√
• Internal systems linked with external ones, such as telephone, microwave, statellite			√

Wiring Closets

Item	Crucial	Important	Unimportant
• Walk-in-shallow		√	
• Walk-in-deep		√	
• Size		√	
• Location		√	
• Expandability/flexibility			√
• Battery room			√

	INTEGRATION WITH SERVICE CORE		
Item	Crucial	Important	Unimportant
• Method of access to floor, conduit, deck			√
• Power requirements			√
• Services (layout)			√
• Service list (voice, data, security)			√
• Fire stops			√
• Wall-mounted			√
• 3 ft deep, access to risers			√
• Separated from sources			√
• Shielded from sources			√
• Patch panels for data systems			√
• Secure			√
• Bullet access to power			√
• Patch panels			√
• Space for multiplexers			√
• Ventilated/power available			√

Wiring Type
Item	Crucial	Important	Unimportant
• Twisted pair (24 gauge)			√
• Caoxtel cable (including Twinax)			√
• Fibre-optic			√
• Dark fibre-optic cable (future use)			√
• Shielded/unshielded			√
• Plenum/nonplenum			√
• Number of conductors			√
• Multimedia cable			√
• Infrared devices			√
• RF			√

Connectors/Wiring Blocks
Item	Crucial	Important	Unimportant
• Single/multiple			√
• Method of connection of wires			√
• Furniture limitations			√
• Orientation			√
• Labelling convention			√
• Termination method			√

Drop Cable
Item	Crucial	Important	Unimportant
• Length limitation			√
• Label convention			√
• Connection to device			√
• Shielded/unshielded			√
• Gauge/size			√
• Transition from wall/floor to device			√
• Undercarpet cable (hot wire)			√
• Number of conductors			√
• Transition from closet			√
• Transition to devices			√

POWER, DATA AND VOICE HORIZONTAL DISTRIBUTION
General Types
Item	Crucial	Important	Unimportant
• Grid density			√
• Palsed access floor			√
• Flat cable distribution			√
• Integrated or cellular floors			√

	INTEGRATION WITH SERVICE CORE		
Item	Crucial	Important	Unimportant
• Flood duct system			√
• Poke-through			√
• Ceiling distribution – hidden			√
• Exposed cable trays			√

Horizontal Modifications
Item	Crucial	Important	Unimportant
• Flexible conduits			√
• Modular removable connectors			√

Workstation Connection
Item	Crucial	Important	Unimportant
• Workstation distribution			√
• Power poles			√
• Raceways			√
• Movable walls/partition distribution			√

Network Topology
Item	Crucial	Important	Unimportant
• Bus (e.g. Apple-Talk)			√
• Ring (e.g. IBM-token ring)			√
• Star (e.g. VAX-Ethernet)			√
• Hybrid			√
• Power voltage/cycle availability at workstation			√

Access to Horizontal Distribution
Item	Crucial	Important	Unimportant
• Composition of floor			√
• Sizes and locations of trenches/cells			√
• Stringers			√
• Height above slab			√
• Pedestal design			√
• Conduit layout			√
• Access to underfloor			√
– access box – configuration (box or tombstone)			
– connection to cabling			
– fire rating/fire stops			
• Smoke detector requirements		√	
• Underfloor distribution from closet			√
– GMD system / point to point			
– wire management - pathways, etc.			
– labelling			
– sealing of floor slab			

Raised Floor
Item	Crucial	Important	Unimportant
• Clear space (inches)			√
• Floor material/module			√
• Pedestal construction			√
• Grounding			√
• Access box/tombstone			√
• Connection to cabling			√
• Fire rating/fire stops			√
• Smoke detector requirements			√
• Grid underfloor distribution from closet			√
• System/point-to-point			√
• Wire management – pathways			√
• Labelling			√

Item	Crucial	Important	Unimportant
INTEGRATION WITH SERVICE CORE			
• Treatment of floor slab			√
Cellular Deck			
• Manufacturer		√	
• Depth and number of cells		√	
• Fire rating/fire stops		√	
• Size and location of trench(es)		√	
• Access to closet/trench		√	
• Access at office		√	
• Access box – configuration and materials		√	
• Connection to cabling		√	
Conduit/Poke-through			
• Conduit layout		√	
• Grid distribution		√	
• Wire management – pathways, etc		√	
• Labelling		√	
Overhead Distribution			
• Power poles offices access		√	
• Wall and column chases		√	
• Grid distribution		√	
• Wire management – pathways, etc		√	
• Labelling		√	
Flat Wire/Flat Cable			
• Number of conductors		√	
• Transition from closet/distribution point		√	
• Transition to devices		√	
Wire Management Systems			
• Diagnostic instrumentation		√	
• Cable identification, wire management software		√	
• Wire identification		√	
• Outlet capability identification		√	
• Wire removal/replacement strategy		√	
• Wiring and telecommunications integration in wireways		√	
• Other wire management software		√	
Planning Module			
• Standard building power for all devices		√	
• Separate feeders and branch circuits for automation equipment.		√	
• Receptacles identified		√	
• Ongoing monitoring of plugged devices into special circuits		√	
• Power conditioning available		√	
Reconfiguration			
• Floors		√	
• 3-D		√	
• Integrated system digital network (ISDN)		√	
• Core distribution		√	
• Punch-down block		√	
Horizontal Distribution/Integration with Other Systems			
• Data, power and telephone distribution	√		
• Data and power distribution	√		

Item	Crucial	Important	Unimportant
INTEGRATION WITH SERVICE CORE			
• Data and telephone distribution			√
• Cable raceways with raised system construction			√
• Cable raceways with ceiling system construction			√
• Cable raceways with furniture system construction			√
• Cable raceways with HVAC duct/pipe distribution			√
• Cable raceways with lighting			√
• Cable raceways and fire management			√
POWER, DATA AND VOICE SERVICE TERMINALS			
Data and Voice Connectors			
• Floor mounted			√
• Wall mounted			√
• Workstation mounted			√
• Single			√
• Duplex or more			√
• Labelling			√
• Length of 'drop' cable			√
• Method of termination			√
Connected			
• Low			√
Transmission Speed			
• Medium			√
• High			√
Connector Type			
• Voice/data			√
• Voice/data/video			√
• Fibre-optic spine			√
• Satellite			√
• Microwave			√
Outlets			
• Type and number of outlets			√
• Integration of outlets into floors, walls, furniture			√
• Labelling of capacity, smart outlets			√
• Modularity, integration with telephone, data, video outlets			√
• Aesthetics, integrity			√
• Safety, security (flush, tombstones, pop-up)			√
Integration of Outlets With Other Systems			
• With furniture			√
• With floor			√
• With walls (movable or fixed)			√
• With ceiling			√
• With HVAC			√
• With lights			√
Appliances			
• Number and density (# per person)			√
• Types (# and density, connected power			√

	INTEGRATION WITH SERVICE CORE
Item	**Crucial** **Important** **Unimportant**

requirements)
- (Desk lamps, fans, clocks, ionisers, radios) — √

Telephone Types
- Dummy — √
- Intelligent — √
- Standard POTS (2500 set) — √
- Feature phones (AT&T, NTI, ROLM, Siemens, NEC) — √

COMPUTERS AND PERIPHERALS

Major Computer Types
- Stand-alone micro (e.g. MACs, other PCs) — √
- Micro with network — √
- Workstations — √
- Minicomputer (e.g. VAX) — √
- Mainframe (e.g. IBM, HP) — √
- Supercomputer (e.g. CRAY) — √

Major Computer Functions
- Word processing — √
- Printing — √
- Electronic mall — √
- Database — √
- Statistical — √
- Accounting, financial — √
- Graphic/CAD — √
- Modelling — √
- Simulation — √
- Publishing — √
- Archival storage — √

Other Data Peripherals (list)
- Line printers — √
- Laser printers (Desktop, IBM 3800, etc) — √
- Plotters — √
- Multiplexer — √
- Network-control equipment — √
- Other — √

Integration of Computer Peripherals and Applicances
- With furniture — √
- With walls — √
- With ceilings — √
- With HVAC/Module — √
- With lighting capability — √

Video Services
- Teleconferencing control facilities (describe setup) — √
- Portable teleconferencing (describe package) — √
- Cable type — √
- Cable vertical distribution — √
- Cable horizontal distribution — √
- Terminal units type — √

	INTEGRATION WITH SERVICE CORE
Item	**Crucial** **Important** **Unimportant**

- Integration w/ power, voice and data — √

Dynamics/Controls
- Power demand and capacity, outlet location, terminal unit location and number, energy management — √
- Telephone (outlet location and number, terminal unit type and number, cost management), power demand — √
- Data (outlet location and number, terminal unit type and number, data management, power demand) — √
- Overload/surge management — √
- Emergency/operation — √

MIS/TELECOMMUNICATIONS DELIVERY PROCESS

During Building Project
- Attendance in general planning meetings — √
- Integral part of building team — √
- Scheduling — √
- Programming and construction drawing review and approval — √
- Budgeting — √
- Right of veto — √
- Integration — √

During Building Operation
- General building user interface — √
- Wire manager — √
- Informed of all moves/adds/changes in MIS/telecom — √

Other Functions
- Service order issuance — √
- Disaster planning — √
- Capacity planning input to building — √
- Programme/occupancy planning — √
- Technology forecasting in building planning — √

NOTABLE OVERALL PERFORMANCE OF POWER, ELECTRONICS, COMMUNICATIONS

Notable Systems Integration
- PEC and structure — √
- PEC and interior — √
- PEC and HVAC — √
- PEC and lighting — √
- PEC and enclosure — √

Lighting Service
- Power capacity — √
- Control flexibility, expansion — √

Fixtures
- Task lighting (table, pole, floor, ceiling) — √
- Ambient (down, up, indirect, daylight) — √
- Task/ambient lighting — √

Item	Crucial	Important	Unimportant
• Reflector, Troffer effectiveness			√
Fixture Type			
• Recessed coffers			√
• Downlights			√
• Track lights			√
• Uplighting, freestanding			√
• Coffer, recessed lighting			√
• Flush tens			√
• Parabolic louvre			√
Planning Module			
• Grid dimensions, densities per sq ft or population			√
• Ease of expansion, substraction			√
• Ease of relocatability, tether/pigtail			√
Lamp Type			
• Cool white fluorescent			√
• Deluxe cool white fluorescent			√
• Warm white fluorescent			√
• High intensity discharge (HID)			√
• Incandescent			√
• A – general service lamp			√
• G – decorative lamp			√
• S – decorative lamp			√
• PAR – lamp used for directional purposes			√
• R – PAR lamp w/t wide beam spread			√
• T – tungsten-halogen lamp			√
• Deluxe mercury			√
• Phosphor-coated metal halide			√
• High-pressure sodium			√
• Low-pressure sodium			√
• Ballasts			√
• Daylighting simulating			√
Central Management/ Control			
• Relay – microprocessor			√
• Electric signals/radio frequencies			√
• Lamp change			√
• Voltage amplitude control			√
• Front-end current limiter			√
• Solid-state ballasts			√
Local Management/Control			
• Automatic shut off, dimming w/time, daylight			√
• Manual/independent switching			√
• Programmable			√
Automatic/Manual Controls			
• Manual			√
• Dimmers			√
• Timer activated			√
• Thermal/heat activated			√
• Motion activated			√
• Sound activated			√

Item	Crucial	Important	Unimportant
• Photoelectric activated			√
Fixture Efficiency			
• Lumens/watt			√
• Watts/sq ft			√
• Reflector efficiency			√
• Bulb efficiency			√
• Lens efficiency			√
• Ceiling/room/furniture configuration efficiency			√
• Surrounding colour, reflectivity and efficiency			√
• Control strategy effectiveness/energy conservation			√
Visual Quality/Performance			
• Appropriate light levels			√
• Light distribution			√
• Glaze control – direct and reflected			√
• Brightness contrast control			√
• Daylight interface			√
• Colour rendition			√
• Fibre optic			√
Spatial Control			
• Multiple switching			√
• Tether			√
• Pigtail			√
• Perimeter/core separation			√
• Number of zones and ease of modification			√
Other Performance Concerns/Opportunities			
• Acoustic/noise generation ballast maintenance			√
• Data/noise interference			
• Acoustic reflection/absorption quality of lenses/fixtures			√
• Thermal/heat generation			√
• Heat recovery/heat minimization techniques			√
• Radiant energy concerns			√
• Outgassing/air quality concerns			√
Integrity			
• Degradation of visual quality control			√
• Degradation of energy effectiveness			√
• Degradation of appearance, discolouration, staining, dirt accumulation, cracking			√
NOTABLE VERTICAL TRANSPORTATION			
Configuration			
• Elevator	√		
• Escalator	√		
• Moving sidewalk		√	
• Minitram (goods, people)		√	
Elevator Type (Operation hrs, connected HP)			
• Single deck			√

	Integration with Service Core		
Item	Crucial	Important	Unimportant
• Double deck			√
• Hydraulic system			√
• Cable system			√
• Gear/gearless			√

Speed

• Low–medium			√
• High		√	

Location

• Central core	√		
• Distributed, multiple core	√		
• External to building	√		
• Staggered cores	√		
• Lower lobby floors	√		
• Sky lobby	√		

Control/Management

• Local, manual or automatic			√
• Central, manual or automatic			√
• Optimisation criteria (ride time, ride frequency, ride capacity)			√
• Speed satisfaction			√
• Comfort, HVAC, lighting quality			√
• Aesthetics			√
• Cost/space effectiveness			√
• Energy effectiveness	√		
• Fire management			√

Emergency Systems

• Power			√
• Air, ventilation	√		
• Light			√
• Phone, safety			√

Integration With Other Systems

• HVAC, spatial effectiveness			√
• Vertical shafts, PLEC			√
• Access, growth potential			√
• Pollution migration control			√
• Pressurisation control			√

SECURITY SYSTEMS
Physical Security

• Lock			√
• Digital password			√
• Card access (magnetic)			√
• Card access			√
• Closed-circuit television			√
• Camera eye/motion detectors			√
• Voice-activated system			√

Employee Identification

• Picture ID			√
• General ID			√
• Guard			√
• Surveillance			√
• Fingerprint/photograph system			√

Employee Access Permission

• Access to building			√
• Access to floors/elevators			√
• Access to specific rooms			√
• Access to computer files			√
• Access to parking			√
• Access to food services			√
• Other			√

Information Security

• Communication scrambling device			√
• Secure telephone system			√
• Nonradiating cable system			√
• Cryptographic devices			√
• IT room security			√
• Cable safeguards			√

Surveillance Type

• Patrols			√
• Motion detectors			√
• Camera (low light, infrared, standard)			√

Intruder Alarm

• Surveillance			√
• Manual 'panic' switch			√
• Barriers/gates			√
• Mantraps/prevention system			√

Security and Building Subsystems

• Enclosure, windows and doors – physical security			√
• Enclosure windows – data security			√
• Interior, workstation enclosures and furniture – physical security			√